DAVE GOULSON

Bee Quest

In Search of Rare Bees

VINTAGE

1 3 5 7 9 10 8 6 4 2

Vintage
20 Vauxhall Bridge Road,
London SW1V 2SA

Vintage is part of the Penguin Random House group of companies
whose addresses can be found at global.penguinrandomhouse.com

Penguin
Random House
UK

First published in Vintage in 2018
First published in hardback by Jonathan Cape in 2017

penguin.co.uk/vintage

A CIP catalogue record for this book is available from the British Library

ISBN 9781784704803

Printed and bound by Clays Ltd, St Ives plc

Penguin Random House is committed to a sustainable future for
our business, our readers and our planet. This book is made
from Forest Stewardship Council® certified paper.

DAVE GOULSON

Dave Goulson studied biology at Oxford University and is now Professor of Biological Sciences at the University of Sussex. He has spent the last twenty years studying bumblebees, and has published over 250 scientific articles on their biology. He founded the Bumblebee Conservation Trust in 2006. He is the author of the *Sunday Times* bestseller, *A Sting in the Tale*, which was shortlisted for the 2013 Samuel Johnson Prize, and *A Buzz in the Meadow*.

ALSO BY DAVE GOULSON

A Sting in the Tale
A Buzz in the Meadow

Contents

For Mum & Dad
Thank you.

BEE QUEST

Prologue

We hiked to the woods, only half a kilometre from the primary school, the kids holding hands in pairs and chattering excitedly. I led the way, toting a selection of beating trays and nets on my shoulder, and their teacher, Mrs Sharkey, fussed and chivvied at the back to keep them all together.

It was a sunny afternoon in June 2009, near the end of the school term, and I was taking the children of my eldest son Finn's class at Newton Primary School, Dunblane, on a bug hunt. Dunblane is a lovely little town, nestled at the western end of the Ochil Hills in Central Scotland, with countryside to be found within a short walk in almost any direction. Once we got to the woods, I handed out the nets and other paraphernalia to the eager seven- and eight-year-old kids and showed them how each worked. All of the nets looked large and clumsy in the children's hands, the butterfly nets being big enough to engulf the smaller children entirely. These kite-shaped nets look easy enough to use, but once a flying insect is caught there is a knack to flicking the end of the net over the frame to trap the creature in a pocket of material and prevent it from flying out again. I showed them how to place a beating tray (a large rectangle of white cloth stretched across a wooden frame) under a low-hanging branch, and then give the branch a good shake to dislodge insects, which tumble, wriggling and scampering in surprise, onto the white sheet. My sweep net demonstration provoked much hilarity – this sturdy white net has to be bashed through long grass, always keeping the mouth of the net facing forwards, which I find is best accomplished by sweeping it from side to side in flowing arcs while stooped forwards, bottom in the air. When doing this I resemble some sort of solo Morris dancer. At the end of the 'dance' I gathered up the bag of the net to prevent the insects escaping, and called the kids round to inspect the catch. Opening up a sweep net is always fun – like opening up a Christmas present, one never knows what marvels will be inside. The kids oohed, aahed and urghed as a myriad tiny creatures – ants, spiders, wasps, beetles, flies and caterpillars – flew, crawled and hopped out of the net. I showed them how to capture the smallest, most delicate ones by sucking them up in a pooter.[1] I dished out a handful of pots each for them to place their captures in, and then the children were off, charging through the under-growth, bashing, sweeping and pooting to their hearts' content, wide-eyed with excitement. We rolled over rotting logs and mossy rocks to find woodlice, ground beetles and millipedes (always carefully rolling them back afterwards). Every new catch was brought back proudly for me to inspect, from huge red slugs to delicate green lacewings. Shrieks of excitement announced the

[1] A small glass jar stoppered with a bung, through which two flexible plastic tubes pass. The end of one tube is pointed at the insects, and the entomologist sucks on the other. All being well, the insect is sucked up the tube and into the jar. One vital feature is that the tube on which one sucks has mesh over the end within the jar – without this, the contents of the jar are liable to shoot up into one's lungs. Even so, it is all too easy in the excitement of spotting an interesting insect to suck on the wrong tube, and inhale all one's previous catches. Delightfully, the pooter was invented in the 1930s by the American entomologist Frederick William Poos, Jr.

capture of a huge queen buff-tailed bumblebee, who buzzed loudly in protest. Finn, bless him, couldn't resist being a bit of a know-it-all and telling the other children what everything was.

It was chaos, but after an hour or so we had a fantastic collection of creepy-crawlies of all shapes and sizes, all laid out in a selection of pots on one of the beating trays. We sorted them into their family groups, learning the difference between flies and wasps, beetles and true bugs, centipedes and millipedes. I told them a little about their diverse and often peculiar lives: which ones ate dung or leaves or other insects; about the parasitoid wasp that eats caterpillars alive from the inside out; and about the froghopper that spends most of its life hiding in a ball of its own spittle. As we let them go, I encouraged the children to hold some of the larger, more robust creatures – there was a beautiful hawthorn shield bug, bright green and rusty brown, with angular, pointed shoulders, which contentedly ambled from hand to hand until, with a flick of its wings, it suddenly whirred away. A half-grown speckled bush cricket, vivid leaf-green with tiny black spots, short-sightedly felt its way along their hands using its outsized antennae, perhaps four times its own body length. A delicate common red damselfly peered at us cautiously with its protruding eyes, as if unable to believe its luck at being released, before helicoptering away on silent, shimmering wings.

As I watched the children's smiling faces, I was reminded of the words of the wise and famous biologist E. O. Wilson, who once said, 'Every kid has a bug period . . . I never grew out of mine.' It is interesting to speculate as to why children are innately fascinated by nature, why they like to collect, be it seashells, feathers, butterflies, pressed flowers, pine cones or bird eggs, and why they love to capture, hold and watch creatures of any and all sorts. I would imagine that in our hunter-gatherer past this curiosity served us well – obviously we needed to build up knowledge of the natural world if we were to survive, particularly as to which animals and plants were dangerous or good to eat, but also so that we could read more subtle signs from nature, interpreting the behaviour of birds that might warn of approaching danger, or indicate the location of water or food. I am often asked where my own early obsession with natural history sprang from, as if I were unusual, but actually I think I was fairly typical – as E. O. Wilson said, almost all of us have a bug period.

The bigger question is why do the large majority of children grow out of their fascination with bugs and, more broadly, with nature? What happens to the child who, aged eight, watched raptly as a woodlouse crawled over her palm? Sadly, by the time they are teenagers, most react to the buzz or scuttle of an insect with a mix of fear and aggression born of ignorance. As like as not, they will swat the poor creature, stamp on it, or at best shoo it away with panicked hands. What goes wrong? Why did their childhood delight evaporate, to be replaced by revulsion? I wonder about those kids in Dunblane, now teenagers. Have they become strangers to insects, forgetting that sunny afternoon, and all that they found so fascinating and fun at the time? Have they absorbed the fears of their parents, the absurd overreaction to a spider dangling from the curtain rail or to a wasp at a family picnic? My family and I have moved from Scotland to Sussex in the south of England since then, but Finn tells me that most of his new friends have not the slightest interest in wildlife – they simply do not see the natural world as in any way relevant to them. Their interests tend to focus on football or PlayStation or posting selfies on Instagram. Without the slightest thought, many of them casually throw drinks cans and crisp packets into the hedge on the way home from school. It is not cool to go birdwatching, and they would think that collecting or photographing or breeding butterflies and moths as a hobby was pretty weird and nerdy.

I would hazard a guess that this change comes about because children get too few opportunities to interact with nature in our modern, urbanised world. Our children will never come to cherish the natural world unless they get to experience it first hand, close up, on a regular basis. They cannot grow to love something that they do not know. If they have never been lucky enough to visit a wildflower meadow in late spring to smell the flowers, hear the bird and insect song and watch the butterflies flit amongst the grass, they probably won't care much if one is destroyed. If they have never had the chance to clamber about in the dappled light of an ancient, wild wood, to kick their feet through the musty leaf litter and emerald leaves of dog's mercury, and breathe in the rich, mushroomy odours of decay and growth, then it will be hard for them to understand what appalling sacrilege it is to rip it down and macerate the trees to make chipboard. Nothing I could write here, even if I had the gifts of Shakespeare himself, could truly convey the wonder and beauty of the natural world. Some fabulous nature documentaries have been made in recent decades, enabling us to marvel at all manner of exotic creatures that we are never likely to see, but I do not think that this is enough, though it may be a good start. We need to get kids outside, on their hands and knees, grubbing about with nature. For me, ten minutes with a bush cricket is worth ten hours of watching a television documentary in which birds of paradise perform their exotic mating dance in some faraway tropical forest.

Sadly, of course, these days few children have the opportunities that E. O. Wilson or I had to allow these interests to develop. More broadly, it seems that children today don't have the chance to explore and experiment in quite the way that was possible for me growing up in the Seventies in a very rural corner of the English countryside. The majority of the world's population now reside in cities – in the UK, a staggering 82 per cent of us now live in urban areas – and children are usually not allowed to roam as they once did. From the age of seven I wandered the countryside around my home village, disappearing off with my friends for hours on end without my parents having any idea where I was. We climbed trees, we fished the lakes and rivers, and we built camps in the woods. These days, of course, young children don't normally get this freedom even if they live in the countryside, for their parents rightly fret over risks from traffic, or less reasonably fear that their child will be abducted by the evil monsters that are imagined to lurk around every corner. It might sound irresponsible of me, but I think children somehow need to be given more chances to explore, to take risks and do foolish, dangerous things from which they can learn. I should know, for during my childhood I did more than my fair share of foolish things, yet I somehow survived.

My very earliest memories involve insects of one sort or another – somehow they burrowed into my soul when I was very young. Aged five, I found the yellow-and-black hooped caterpillars of the cinnabar moth feeding on the groundsel growing through the cracks of my primary-school playground, and packed far too many of them among the crumbs left in my lunchbox to take home. I collected more groundsel to feed them, and was thrilled when some of them eventually developed into adult moths, weak-flying but beautiful creatures with glossy magenta and black wings (which I learned much later are a warning that they are poisonous, having accumulated the toxins that are supposed to protect the ragwort from being eaten). I collected millipedes, woodlice and beetles from the garden, and the tiny red mites that scurried about on the low concrete wall in front of our house on sunny days, and I kept them all in jam jars, lined up on the windowsill of my bedroom. I guess many of the poor creatures died, but I learned a huge amount, not least from the *Oxford Book of Insects* that my parents bought me so that I could find out what my catches were. In the evenings, I pored over the watercolour illustrations and made plans for local expeditions in which I imagined that I might find some of the more exotic creatures – great silver water beetles, emperor dragonflies and death's head hawk moths.

When I was seven we moved from our small semi-detached house on the edge of Birmingham to the rural village of Edgmond in Shropshire, which provided many more opportunities for creature-hunting. I made school friends who were similarly minded, and we would spend our lunchtimes searching the hawthorn hedges along the edge of the school field for the beautiful caterpillars of the yellow-tail moth, velvet black adorned with a crazy Mohican row of red, black and white tufts of hairs. At weekends we searched for other types of caterpillar, scouring the hedgerows, meadows and copses around our village. With the help of the *Observer's Book of Caterpillars*, another gift from my parents, we worked out what each type was as best we could, and found the correct leaves to feed them. I found their specificity intriguing – most moth and butterfly caterpillars will eat just one or perhaps two types of leaf, and will simply starve to death rather than try to eat anything else. A few types are much less fussy – the enormously hairy black and orange caterpillars of the garden tiger would eat almost anything apart from grass.[2] On one occasion we found a caterpillar of the puss moth feeding on a willow, a fantastic green and black creature which reared up when frightened, and extruded a pair of intimidating red, writhing tentacles from its forked tail. I had to wait nearly a whole year until the following spring before I got to see the adult moth: a splendid, fat-bodied, furry kitten-like animal, its body and wings snow-white speckled with black.

[2] The caterpillars of the garden tiger used to be very common across England, and were well-known to children as woolly bears. Few children today will have seen one, as this species has undergone a massive decline during my lifetime.

When I was only seven or eight I began collecting birds' eggs, something my dad had done himself as a boy. As I recall, almost every boy in my village had a collection (I've no idea what the girls did – having no sisters and going to a boys' grammar school, until the age of about fourteen I was almost completely unaware that girls existed). We vied with each other to find the nests of the more unusual species, and coveted each other's finds. Once again, the *Observer's* series of natural history books was invaluable – I still have my tattered copy of the *Observer's Book of Birds' Eggs*, nearly fifty years old. I remember finding a blue egg with pale brown speckles lying abandoned on the ground on the slopes of the Long Mynd in south Shropshire, and I convinced myself that it was the egg of a ring ouzel, a spectacularly rare moorland bird that I had never actually seen. My friends were sceptical, and we argued about its identity for days, though with hindsight I'm pretty sure it was just the egg of a blackbird. We learned a huge amount about the natural history of birds in the process, for each species tends to nest in particular places, makes its nest from characteristic materials, and so on. On a couple of occasions we found nests of the long-tailed tit, extraordinarily beautiful spherical constructions woven from spider's web and soft moss.

I moved on from this to collecting butterflies, and expanded this to moths, and then beetles, and eventually became fairly expert in identifying them all. My skills in rearing moths and butterflies came in useful, for they enabled me to get perfect, untarnished adults for my collection, but by the age of twelve or so I eventually tired of killing these lovely creatures and began rearing them simply to release back into the wild. In particular, I reared hundreds of peacocks and small tortoiseshells, collecting up the young caterpillars from nettle patches and rearing them up in home-made cages where they could not be attacked by the tachinid flies and chalcid wasps that parasitise most of them in the wild. It was a heart-warming experience watching the young adult butterflies tentatively take flight for the first time, their pristine wings freshly dried, fluttering upwards and eventually soaring out of our garden.

It wasn't just natural history that captured my youthful attention. When I started secondary school, I quickly came to love science of all types, particularly the pyrotechnics of chemistry and the thrill of danger associated with electricity. My parents gave my elder brother, Chris, and me a chemistry set and, as countless children have before and since, we spent hours heating up random mixtures of chemicals on the small methylated-spirit burner, usually creating nothing but a sticky brown mess and a cloud of noxious smoke. Risking an instant detention or worse, my friends and I would smuggle small pieces of magnesium ribbon out of chemistry classes and delight in setting light to them in the bushes at the bottom of the school playing field during lunch break. They burned so brightly that we had white spots before our eyes through the afternoon lessons. After seeing demonstrations in which our teacher dropped small pieces of sodium or potassium into a sink of water – at which point these highly unstable metals fizzed and banged, sending up spurts of flame and clouds of steam – we longed to get our hands on some, but our unsporting teacher never let them out of his sight and always locked them away in a metal cupboard at the end of lessons.

Luckily my parents were as tolerant of my early chemistry experiments as they were of my enthusiasm for filling the house with jam jars, cages and tanks full of creatures, although they rarely knew exactly what my friends and I were up to. As we learned a little chemistry, we managed to devise ways to conduct ever-more dangerous and entertaining experiments at home. With my friend Dave (there were five Daves in my class at school, and for boys of my generation it might have been useful if someone had invented a collective noun for groups of us) we worked out how to produce hydrogen and oxygen gas by passing electricity through water. The transformer from my Scalextric set proved to be an invaluable power supply for such experiments, producing a steady twelve volts that was ideal for this. The hydrogen and oxygen could be collected in bottles and the two gases together exploded with deliciously satisfying violence when lit with a match, though the lighting was not without minor risk. I even learned to produce chlorine gas in a complicated experiment on the kitchen worktop that involved passing electricity through domestic bleach; the clouds of brown gas are highly toxic, and the experiment was so unexpectedly successful that I nearly expired before I could turn the apparatus off and get the windows open.

At around this time Chris and I were collecting second-hand books to sell on a stall at an upcoming school fete – although I can't recall, I imagine that my father volunteered us for this task as I can't imagine my brother or I ever volunteering to go door-to-door in the village with a wheelbarrow, collecting unwanted books. However, as it turned out, it had unexpected benefits; amongst the piles of yellowing romance novels and endless Agatha Christie murder mysteries, I found a small book called simply *Explosives*. You can imagine my excitement at opening this treasure and discovering it contained details of how to make a multitude of highly dangerous, sometimes unstable compounds. It was disappointing that most of the instructions required reagents that were unavailable to a boy of twelve; for example, it was clear from the outset that I would never find a way to get hold of the necessary quantities of concentrated acids required to make TNT. However, the recipe for gunpowder looked tantalisingly possible. Gunpowder, or black powder as it is cryptically known to aficionados, contains just three ingredients: sulphur, charcoal and potassium nitrate. My children's chemistry set contained sulphur, and charcoal was easy enough, although grinding barbecue charcoal to the required powder was a messy process. That left only potassium nitrate. The book explained that there are significant quantities of potassium nitrate in pigeon excrement, and that extraction was possible with care. It took some time to locate a pigeon fancier in the village, but eventually, by dint of a lot of furtive peering over garden fences, we spotted a pigeon loft complete with cooing occupants. If we had possessed any common sense, we would have simply knocked on the door of the house and asked for some pigeon droppings – I suspect that the owner would have been happy to give us some, so long as we gave a vaguely plausible explanation – but we feared that the owner might rumble our true intentions. Of course, with the benefit of hindsight, it seems unlikely that they would have jumped to the conclusion that we wanted the droppings so that we could make a bomb, but in our paranoid imagination this seemed a real possibility. Anyway, once we had decided against the direct approach a covert night-time operation seemed the obvious alternative. My mate Dave (one of the many) and I snuck into the garden one dark evening and were relieved to find the loft unlocked – pigeon rustling and dung theft presumably being rare occurrences in rural Shropshire at the time. It was messy and extremely smelly work scraping up the droppings into a carrier bag in the pitch black – we dared not turn on a torch – and the pigeons started to make a racket, nervously flapping about and splattering us with droppings from above, so we made a hasty retreat, satisfied with our haul. I have often wondered if the pigeon fancier noticed that someone had mysteriously cleaned out his pigeons in the middle of the night.

The next day we set about extracting the potassium nitrate. The book didn't explain how to do this, which was an unfortunate oversight on the part of the author. We knew it to be water soluble, so using our rudimentary knowledge of chemistry we figured that we should be able to rinse the potassium nitrate from the faeces,

sieve out the solids, and then extract the chemical from the resulting solution. At the bottom of my garden we mixed the
droppings into a bucket of warm water, and then sieved out the
lumps using an old tea towel. It was pretty unpleasant work. We
ended up with a bucket of extremely smelly, pale brownish liquid.
We decided that all we then needed to do was drive off the water
by boiling the liquid for a while, which should hopefully leave us
with something that was mostly potassium nitrate. I started this
process in an old pan on the kitchen stove, but predictably and
understandably my mum immediately evicted us from the house.
Luckily I had previously rigged up a Bunsen burner to an old
camping-gas cylinder in the shed, so we resorted to that. It took
hours, and as the liquid thickened the stench became ever more
horrendous, but eventually the pan's contents had boiled down to
a sticky brown mess. It didn't look much like potassium nitrate,
which we knew was supposed to be a white crystalline solid, but
we hoped that it might do the job.

We carefully mixed the brown goo with sulphur and charcoal
in the allotted proportions. The resulting mess was an interesting
greenish black paste. We took a small portion, placed it on the
bottom of an upturned tin can, and I gingerly applied a match,
my heart hammering with excitement. The match stuttered, the
powder spluttered, and then . . . nothing. I tried again and again
but it was hopeless. Clearly there was not as much potassium
nitrate in pigeon dung as we had hoped, or perhaps our extraction
method was ineffective, or maybe they were just the wrong type
of pigeons.

A little research revealed that potassium nitrate was sometimes
sold as a garden fertiliser. In fact, a small gardening shop close to
my school in Newport turned out to stock it, along with a range
of other desirable chemicals, but all were kept on a high shelf
behind the counter. My friends and I surreptitiously checked them

out while pretending to browse the packs of vegetable seeds. Eventually I plucked up the courage to try to buy some, certain that the shopkeeper would suspect my true purpose. He was an elderly, grey-haired man with a stern air, and he immediately started quizzing me as to what I wanted it for. I went bright red with embarrassment – I've always been a hopelessly unconvincing liar – and stammered that it was for an experiment for school, to see how potassium nitrate affected how well plants grew. My friends had formed a phalanx behind me for moral support, and the bolder ones chipped in various additions, including something about a school competition to see who could grow the biggest vegetables. It was vaguely plausible, though unlikely, but I stubbornly stuck to my guns as he cross-questioned me, and eventually he reluctantly brought down a two-pound box from the shelf. I'm sure he knew we were up to no good, but he couldn't prove it and perhaps he was glad to sell something, for the shop was always very quiet. I handed over my money, grabbed the box, and we sped off before he could change his mind.

Gunpowder proved to be tremendous fun. It didn't explode but it burned ferociously, emitting clouds of sulphurous smoke, the evocative smell of fireworks on a cold November night. We experimented with different proportions of the ingredients, setting off small piles on a piece of slate at the bottom of the garden where the prying eyes of parents were unlikely to notice us. As we honed the mixture it burned ever faster and lighting it with a match often resulted in singed fingers, so we worked out how to make fuses from twists of loo paper soaked in potassium-nitrate solution and then dried out. We experimented with adding other chemicals from our chemistry sets to try to change the colours of the flame or the smoke, and we packed tubes of cardboard with gunpowder with various additions to produce our own primitive fireworks. They were all pretty hopeless compared to professional fireworks, but as with all things home-made they were somehow a lot more satisfying than the bought variety.

My friend Dave came up with an alternative pyrotechnic formula, based on sodium chlorate weedkiller mixed with sugar, and we vied with each other to produce the best fireworks. We spent weeks trying to create rockets that would actually take off, though we never got the hang of this – the highest we ever managed to get one to go was about four feet into the air, before it flipped over and hurtled to the ground. Our garden lawns became pocked with brown scorch-marks from our many failed launch attempts.

Although the powders we created were highly inflammable, they did not actually explode, which was something of a disappointment. Eventually we discovered that the only way to create explosions was to seal the powder inside a more or less airtight container, and then light it. This of course is tricky, for how do you light something once you have sealed it in a container, and how do you do so while maintaining a safe distance so as not to get blown up? My book, *Explosives*, was little help on this point. After a lot of discussions, trials and errors, Dave and I found the answer, in the form of the old-fashioned disposable flash cubes used in photography. Younger readers may be surprised to hear that it wasn't so long ago that cameras didn't come with a flash as standard, but instead had a mounting point for a disposable plastic flash cube which contained four one-use bulbs. Each time you took a picture, the forward-pointing bulb would burn white hot and self-destruct, producing enough light to take a single photograph. You would then rotate the cube a quarter-turn to ready the next bulb for action. Amazingly, the only power needed to make one of these bulbs fry itself was a normal 1.5 volt AA battery.

We found that these bulbs, once carefully dissected out of the plastic casing, would readily light my gunpowder or Dave's weedkiller mix. So we made thick tubes of cardboard and filled them with our pyrotechnic powder along with a flash bulb attached to two thin wires that led out of the tube. We sealed the tubes up with layers and layers of gaffer tape. All we then needed to do was attach the wires to the terminals of a battery and, hey presto, BANG! The tubes would fly apart with an ear-splitting crack, leaving only a few smoking remains. It was brilliant fun, and before long we moved on to using copper pipes to get a bigger bang – these really made the earth shake when they went off, leaving scraps of twisted metal lying around. To ensure that we were at a safe distance, we rigged up the battery with an old-fashioned alarm clock, with a wire pushed through a hole drilled in the glass face making contact with the minute hand when it reached the vertical. In this way we could set the bombs off with a delay of up to about fifty-five minutes, and then sit and watch them go off on cue from a few hundred yards away. We had lots of fun with these home-made pipe bombs, planting them in holes in trees, crevices in the rock wall of a local abandoned quarry, and once in holes in the brick wall of a crumbling, abandoned farm building. They weren't particularly powerful, but would usually blast a few bits of wood or rock or brick up into the air. On one occasion we even put one in the local canal, having seen dynamite fishing on the television. The blast didn't kill any fish, but it produced a satisfying eruption of water.

Bomb making might not seem the safest of activities for young teenagers to be engaged in, and I would absolutely not encourage such things, but it was relatively harmless compared to our tampering with the local electricity supply. On one ill-fated Sunday morning when I was thirteen, my friends Matt and Tug (Tim) and I were messing around in my garden with a piece of old, rusty barbed wire we had acquired from somewhere. It was a couple of yards long, and made an interesting whistling noise when whirled around enthusiastically above one's head. However, this didn't keep us interested for long and so for some reason I decided to whirl it round and then attempt to throw it from the garden, across the road in front of our house, and into the field beyond. I hadn't noticed the electrical cables strung from telegraph post to telegraph post along the street. The barbed wire hit one, snagged, and swung around to contact a second cable at which point there was a loud bang, a shower of orange sparks, and two pieces of barbed wire fell to the ground. On closer inspection we discovered that the wire had melted right through in the middle, and was still glowing red hot on the pavement. Presumably the high voltage electricity shorting through the barbed wire had been too much for it. This was brilliant sport, and of course we wanted to do it again.

It dawned on us that it might be wise to find somewhere a little more secluded – my front lawn not being the most discreet of locations. So we wandered off towards the edge of the village, searching for some more barbed wire as we went, since the pieces we now had were too short. It took a while to find any, but eventually we found an old coil of surplus wire attached to a fence post in the corner of a field, and by dint of a lot of bending the wire backwards and forwards we managed to break off a piece. We took it with us and headed off up the nearest lane, beyond the last house, until we found some more overhead cables. With hindsight, we should perhaps have noticed that these cables were higher than the ones outside my house, and mulled upon the significance of that fact. We ought also to have noticed that they were somewhat thicker, but what with them being so high this wasn't all that obvious. Regardless of such subtleties, we set about hoiking our piece of barbed wire at the cables. Because of their height, this was much more difficult than with the cables by my house. We took it in turns, whirling the wire around our heads and launching it skywards. Every now and then the wire would hit one cable and fall back without effect. It took us nearly two hours before, as luck would have it, it was I that finally managed to get the barbed wire to snag one cable, spin, and touch a second. What happened next is indelibly etched into my memory. There was a deafening bang and a white flash that resembled lightning. One of us shouted 'RUN!' – it may have been me, or it may have been all of us simultaneously. We fled. As we pelted towards the village, I glanced back to see the two overhead cables fall to the ground, thrashing and sparking as they did so. This was not quite what we had intended.

We ran back to my house, which happened to be the nearest, and we hid in the garden shed. We sat on the piles of second-hand romance novels left over from the school fete, pondering our next move. We knew this was bad, and could not see much chance of escaping major trouble. We'd spent so long in the lane by the power lines that at least a dozen cars had passed by that morning, and in our small village everybody knew everybody; it wouldn't take long for someone to work out who the culprits were. Eventually we decided that there was nothing for it but for each of us to go home and confess. With my stomach in my boots I walked in through the back door of our house to find my mum in an uncharacteristically bad mood. She'd been in the middle of roasting a big joint for Sunday lunch, and there had been a power cut. There was no gas in the village, so every Sunday lunch had been cooking in an electric oven. Now, all over the village, half-cooked chickens and sides of beef were slowly cooling. In the two village pubs, the Lion and the Lamb, dozens of Sunday lunches would never now be properly cooked. In the late 1970s power cuts were quite common, but they were usually at night and there was usually a warning in advance. On this occasion, of course, there had been no warning.

This was something I hadn't anticipated, and I ran outside again without saying a word to my mum; Tug and Matt were still within sight, since each had been walking rather reluctantly and hence slowly in opposite directions towards their respective homes. I called them back and told them what had happened. It was much worse than bad. This was a disaster on a biblical scale. We hid back in the shed. Matt suggested, without conviction, that perhaps the power cut was a coincidence. We knew it wasn't. In fact, as it subsequently turned out, we had by chance hit upon the 11,000-volt power lines that were the sole power source to the village. It took most of the rest of the day for an emergency team from the electricity board to repair them. My friends and I were still sitting in the gloom of the shed when the local policeman arrived in his Mini police car. He was not particularly well disposed to us since a couple of years earlier he had caught us taking potshots at his geese with our home-made catapults (which he had confiscated and incinerated), so he delighted in carting us off to the tiny police station in nearby Newport.

In the end, we got off with a small fine and slapped wrists. The worst of it for me was the embarrassment caused to my dad, who as a local schoolteacher saw himself as a pillar of the community. Naturally enough, he was mortified to have his son hauled in front of the magistrates. To make matters worse, the headmaster of his school also lived in our village and had lost his Sunday lunch on that fateful day.

Of course I'm not advocating that children be allowed to go around blowing up farm buildings and sabotaging power lines, or for that matter collecting birds' eggs. Some of the many things that we did were highly dangerous and idiotic. However, I am not sure that I would have become a scientist as an adult if I had not been able to indulge at least some of these youthful enthusiasms in the ways that I did. Perhaps my parents were too tolerant, and probably also somewhat naïve, but I am enormously grateful that they gave me as much slack as they did (though perhaps some words of wisdom about the dangers of high-voltage electricity might have come in handy). I try to let my own boys, now five, twelve and fourteen years old, have enough freedom to learn for themselves. I wince when I see them swinging from branches near the tops of tall trees, and perhaps I shouldn't let the five-year-old play with my axe or my hammer drill, but at the time of writing they have so far all survived. I've bought them ingredients for home-made fireworks, although I try to keep an eye on what they are up to and have ruled out pipe bombs as a step too far. They too haven't yet got a rocket to take off, and our lawn bears numerous scorch-marks from their unsuccessful launch attempts. I've also tried to give them every chance to engage with the natural world. We are lucky, for we live in the weald of rural Sussex, surrounded by woods, pasture and streams, which they can explore in relative safety – the biggest dangers they are likely to encounter are themselves. In the summer, we go down to our little farm in the deepest, darkest French countryside where they can run amok. I don't know whether they will follow me in studying natural history, but at least they have had ample opportunity to fall in love with nature. My eldest, Finn, can now identify most wildflowers, and Jedd has become an adept insect photographer. Seth, the youngest, simply wants to catch everything, put it in a Tupperware container, and watch it – he is very much still in his bug period, long may it last. I am sure that they will do their best to champion nature's cause in the future.

Sadly, I fear that they are the exception. I do not know for sure, but my impression is that engagement with nature is declining, that the generation growing up today are even more detached from the world that supports them than the one before, and if so then it is a terrible thing. Even today, in the midst of a mass extinction event caused solely by man's activities, with climate change threatening to render large portions of the globe near uninhabitable in the not-to-distant future, and with topsoil being lost at the rate of about one hundred billion tonnes per year, environmental issues remain pretty low on the political agenda. The environment was scarcely mentioned in the 2015 UK general election campaigns, even by the Green Party. Most of the debate focused on the economy, but money will be little use to us when we have no soil or bees.

If we want to save the natural world, and ultimately to save ourselves, then we need more people to care about its fate. First and foremost, we need to ensure that our children grow up with opportunities to explore nature for themselves, to get covered in mud chasing frogs or crawl through hedges looking for caterpillars. We need to give them the opportunity to express their natural curiosity, to watch a butterfly emerge from her pupae, to see tadpoles developing tiny limbs, to experience the excitement of discovering a slow-worm under a log. If we give them this, then they will love nature, cherish it and fight for it in the future.

I was fortunate enough to do all of these things as a child, and it inspired me to spend the rest of my life pursuing my own curiosity with regard to natural history. I have been lucky enough to travel the world and have watched birdwing butterflies soaring through the rainforests of Borneo and listened to howler monkeys proclaiming their territories in the forests of Belize, amongst so many other unforgettable experiences. Much closer to home, I have spent countless happy hours hunting for insects, birds, reptiles, mammals and flowers in the less spectacular but just as wonderful woods and meadows of France and Britain. I have been lucky – I was brought up in the countryside, and then stumbled into a career that allows me to spend my time chasing after the world's most interesting bees, in the hope of understanding more about them, of unravelling some of the unknown details of their lives, and trying to work out how we can conserve them so that others might enjoy them in the future. This book is the story of those bee travels. We'll start close to home, in some of the hidden corners of Britain where wildlife still thrives, before moving abroad to the wild mountains of Poland and then to the Andes and Rocky Mountains of the New World, where a sad tale is inexorably unfolding for their bumblebees. Finally, we will return to Britain for some inspiring and hope-filled examples of nature's resilience. Welcome to *Bee Quest* . . .

CHAPTER ONE

Salisbury Plain and the Shrill Carder

Somewhere, something incredible is waiting to be known.

Carl Sagan

I have in the past been heard to blame Adolf Hitler for causing the demise of British bumblebees, for it was the Second World War that really started the drive to increase food production in the UK – at the time, it made sense for Britain to attempt to become as self-sufficient as possible for food, as there were precious few routes for importation. Thus began decades of agricultural intensification, during which a good deal of our countryside was destroyed to make way for vast monocultures of crops. However, if I continue with this line of logic then I must concede that the Kaiser and Hitler perhaps deserve some grudging acknowledgement, for their actions also unwittingly led to the creation of one of the largest nature reserves in Europe.

In 1897 the Ministry of Defence began purchasing land on Salisbury Plain on which to conduct military training exercises.[3] At the time, Britain had a vast empire and had been involved in a long succession of conflicts around the globe – it was a lot of

[3] Chris Corden's *Salisbury Plain: Military and Civilian Life on The Plain since the 1890s* provides an engaging account of the army's varied and sometimes eccentric activities on the Plain through to the present day.

1

work claiming new territories in far-flung corners of the Earth, and it took a lot of well-trained troops to keep so many poorly armed native peoples firmly under our thumb. During the sixty-three years of Queen Victoria's long reign, we were involved in no less than thirty-six full-blown wars, plus eighteen military campaigns and ninety-eight military expeditions. Our standing army was vast, and we needed somewhere to train all those men. Recognising this, the government passed an act enabling the army to buy its own land, by compulsory purchase if necessary. It made sense for the army to focus on an area not too far from transport links, London and the Channel ports, somewhere where there were few people, and where land prices were low. Salisbury Plain fitted the bill perfectly, for the collapse of the wool industry in the mid-1800s had made Wiltshire one of the poorest counties in Britain. The army began an extended shopping spree – in 1897 alone it bought about 6,000 hectares of the Plain, plus various other chunks elsewhere in Britain.

Prior to the arrival of the army, the Plain had had an ancient history of human occupation. It comprises a huge slab of chalk, laid down as the shells of countless trillions of tiny dead sea creatures accumulating at the bottom of an ancient sea some hundred million years ago, but now raised up in a rolling plateau that gently inclines from south to north, and reaches not much more than 200 metres above sea level at the highest points. It would have become forested as the ice retreated from Britain after the last ice age, but along with the North and South Downs it was one of the first areas to be cleared of trees by early Neolithic settlers perhaps 5,500 years ago – the thin chalk soils would have made it less difficult to grub out the roots than in the lower surrounding areas. There are signs of human activity from even earlier – the rotted stumps of a line of regularly spaced upright poles sunk into the ground some 8,000 years ago, for an unknowable purpose. We can glean very little about how these people lived, but their presence is evidenced by the many strange barrows, tumuli, hill forts and other odd-shaped mounds of mysterious origin that are littered across the Plain.

Of course, the most famous Neolithic structure is Stonehenge, a wonderful and iconic collection of enormous cut stones, a circle of vertical pillars, the Sarsen stones, capped by massive cross-pieces, somehow erected 5,000 years ago. I visited Stonehenge when I was just a child, at a time when visitors were allowed to walk and clamber amongst the stones, and the memory remains vivid to this day. There is an inescapable magic to this cluster of ancient stones, which some fancifully claim were carried there and erected by the legendary wizard Merlin. Certainly it is hard to explain how they were transported there by conventional means, for the four-tonne 'blue stones' which make up a smaller, inner circle at Stonehenge were quarried in west Wales, some 290 kilometres distant, with major rivers and mountains in between. If no magic was involved then one imagines that there must have been an awful lot of blood, sweat and tears expended instead, so presumably these people thought that building Stonehenge was pretty darned important. The Sarsen stones originated from near Avebury, a mere forty kilometres to the north, but weighing twenty tonnes each must also have taken some shifting. It has been calculated that it would take a team of about 600 men to drag each stone on rollers, and even with so many it would have been very, very slow. One imagines that assembling such a team must have been quite a task, at a time when the total population of the UK might have numbered just a few tens of thousands. Why these ancient people went to such extraordinary lengths remains entirely unknown. Cremated human bones and other remains have been found buried in pits at the site, and radio isotopes have revealed that some of these remains are of people who originated far away, in France, Germany, even a boy from the Mediterranean basin. Perhaps these were human sacrifices, foreign slaves slaughtered to appease a long-forgotten god? Alternative theories suggest that the stones were a site of astronomical study, or a place of healing, even a celebration of peace and unity amongst the Neolithic peoples. We'll almost certainly never know. Whatever their original purpose, it feels as if these past activities have somehow left their mark, for there is no doubt that the stones have a brooding atmosphere all of their own.

Much later, the Romans arrived and grew crops on the Plain to feed their legions. Later still, in the year 878, King Alfred is thought to have won a decisive battle against the Viking invaders near Westbury, the victory commemorated by a white horse carved into the chalk of the hillside above Westbury on the western edge of the Plain. Through all of this until the beginning of the twentieth century, it is likely that the lives of the people who lived and worked on the Plain changed relatively little. A pair of Victorian sisters, Ella and Dora Noyes, travelled the Plain in the late 1800s and published an evocative illustrated account of their experiences in 1913: *Salisbury Plain: Its Stones, Cathedral, City, Villages and Folk*. Life on the Plain was centred on sheep – shepherds each maintained flocks a thousand strong, with wool being the main source of income, the animals providing not only meat but also fertiliser for the arable fields. The villages on the Plain tend to be tucked into the valleys for shelter and surrounded by enclosed fields, with the surrounding plain comprising mostly open pasture. These pastures had probably been managed in the same way, by occasional livestock grazing, for the best part of 5,000 years. Ella Noyes described the village of Imber:

The village lies in a deep fold of the Plain, on the track of another little winter stream; on all sides the slopes of the high downs surround it. It is just one straggling street of old cottages and farmsteads, winding along the hollow under the sheltering elms; the narrow stream brims fresh and clear through it in spring, leaving its bed dry, to fill up with coarse grass and weeds, in summer.

The white-washed cottages, with their leaning timbers and deep thatched roofs, are set down in short rows and groups, the angles and nooks between them filled in with garden plots full of flowers; rose bushes, here and there a lilac, lilies, and tangles of everlasting peas.

There is known to have been a settlement at Imber since at least 967, and the village is recorded in the Domesday Book one hundred years later. When the Noyes sisters visited there were small shops, a pub, a blacksmith, a windmill to mill the grain for bread, a small school, a Baptist chapel and a substantial church. Their description of the village and the people that lived there make it sound romantic and idyllic, but life must have been blisteringly hard. Most folk would leave school and go to work at the age of nine; those that were not shepherds were farm labourers, maids, blacksmiths, millers or bakers – hard manual jobs that must have changed little over the centuries. The Noyes sisters cannot have known it, but they were describing a way of life that was just about to disappear.

In 1898 over 50,000 troops took part in exercises and marching parades on the Plain in preparation for the Second Boer War in South Africa, a sign of things to come. There not being too many sources of public entertainment at the time, these military man-oeuvres became a popular attraction, with hundreds of local people turning out at the weekend to picnic on higher points of the Plain and watch the mock battles. More permanent changes began with the onset of the First World War, for large numbers of overseas troops, particularly Canadians and Australians, were stationed on the Plain, training alongside home-grown volunteers. Most had to live under canvas, or at best were allocated a bunk in primitive wooden barracks that were hastily thrown up. The sleepy villages were suddenly choked with soldiers, horses and wagons, and businesses of all sorts sprang up, from barbers to brothels.

During the war, early experiments with military aircraft took place on the Plain, with quite a few crashes and several lives lost. At one stage the fledgling Royal Air Force considered moving Stonehenge to make way for a runway, but luckily other suitable sites were located. At around the same time, an engineering company based in the village of Bratton was asked to develop a new, top-secret, all-terrain metal war vehicle carried on caterpillar tracks. Since the huge and noisy prototypes could hardly be hidden, the company created a cover story that they were building a machine to carry tanks up onto the plain to water the sheep. It is plausibly claimed that this is how the name 'tank' came about. One imagines that the local folk must have had much to talk about of an evening in the village pubs.

The winter of 1914–5 was particularly wet, with extensive flooding in the valleys of the Plain, and many of the soldiers perished of diseases such as meningitis before ever making it over the Channel. One imagines that conditions in the crowded, muddy military camps would have provided good training for the horrors that lay ahead on the Western Front. For many men, their last memories of England are likely to have been of their soggy months of training on the Plain before being shipped over to France to be slaughtered like cattle. Of those that were lucky enough to survive and return to England after the war, many were initially returned to their billets on the Plain, where the Spanish Flu pandemic took yet more lives. There are military cemeteries near most of the army camps on the Plain, some of the graves occupied by those who eventually succumbed to wounds received in battle, but many more occupied by those who perished from disease.

After the First World War military spending was cut, for the prospect of another war in the near future was unthinkable. For a brief period the army had difficulty making ends meet – it even resorted to getting soldiers to hunt the innumerable rabbits on the Plain for selling to local butchers to raise money. The military presence on the Plain dwindled for a little while, and life would have returned to something approximating normality for many of the locals, but of course it was not to last.

As unrest in Europe began to grow once more, the army recommenced its programme of land purchase, acquiring the village of Imber in 1927 (all except for the church). Until the Second World War life continued in the village and elsewhere on the Plain much as it always had, despite the change in landlord, but the war brought a new influx of soldiers. In 1943 the Imber villagers were evicted so that their homes could be occupied by American troops, and so that troops could practise for the D-day landings in secret. The villagers were initially told that their furniture could be put in storage, and that they would be allowed to return to their homes after the war was over, but they never were. With just a few weeks' notice, the farmers were forced to sell their herds at knock-down prices – over 5,000 sheep and seventy dairy cows – and their farming machinery.

Such heavy-handed treatment of the local people is of course awful, but the sequestration of all this land did, at least, protect it from agricultural change. Elsewhere in Britain, the drive for self-sufficiency during the war coincided with the increased mechanisation of farming, along with the advent of cheap artificial fertilisers and the introduction of synthetic chemical pesticides, and together these innovations fuelled dramatic changes in farming practices. The end result is the vast monocultures of chemical-soaked crops that constitute food production in the modern world, and catastrophic crashes in wildlife populations. Almost all of the flower-rich chalk grasslands of the North and South Downs in the south and east of England were ploughed up to create arable fields or 'improved' pastures, but on Salisbury Plain much of the original downland remains to this day. Whatever one's opinion on the morality of the army's actions in evicting people from their homes and the compulsory purchase of land, an unintended consequence was the creation of a vast unofficial nature reserve. Today, the army's holdings on the Plain comprise about 400 square kilometres of land – or 40,000 hectares. This isn't the entirety of the Plain by any means, but it is more than half – a seriously large patch of land by any standards.

After my childhood visit to Stonehenge, I was not to return to the Plain until the winter of 2002. At the time I was a youngish lecturer at Southampton University, which is about forty kilometres to the south, so it was not a huge journey. I'd been studying the behaviour and habits of common bumblebees for six or seven years in and around Southampton, and I was struck by the fact that I had never seen many of the UK's species. Even the ruderal bumblebee and brown-banded carder, which according to distribution maps published in the early 1980s were species that ought to be found in the south of England, did not seem to occur in south Hampshire any more. I had heard that Salisbury Plain supported populations of many rare insects and flowers, and it seemed to be the most likely place within striking distance where I might encounter some of these rare and exotic-sounding bumblebees. So it was that one dull morning in February I drove up to the army barracks at Tisbury for a compulsory safety briefing on accessing the Plain.

The briefing was delivered by a short, barrel-chested sergeant with a majestic moustache, almost a comic-book caricature of an army type. He sternly described the various risks one might encounter, making it sound as if the odds of surviving a visit to the Plain were slim indeed. He explained that there is substantial unexploded ordnance all across the Plain, the accumulated debris from over a hundred years of military exercises, so that sticking to the main paths is advisable, and digging in the ground or picking up metal items is a very bad idea and strictly forbidden. Access to the central artillery bombardment area is off-limits at all times, as you might expect, but live firing also takes place in other areas, warning of which is given via a series of flags. He made it abundantly clear that, when driving on the tracks across the Plain, it is wise to give priority to Challenger tanks which, weighing over sixty tonnes and travelling at up to sixty miles an hour, can go right over a domestic car and barely notice. This seemed like sound advice.

A few months later, on a cool day in early June, I returned to hunt for bees. I was driving my slightly silly two-seater sports car, a black Toyota MR2, definitely not a vehicle suited for a head-to-head encounter with a tank. I passed through the town of Bulford, dominated by its huge army barracks, and headed along a narrow lane which wound northwards and quickly degenerated into an unsurfaced and heavily rutted track to which my car was extremely poorly suited. I passed a warning sign telling me that I was entering the military training area, but thankfully there was no red flag flying to indicate that I was likely to be blown up or strafed by rifle fire in the imminent future. The track climbed gently, and after another quarter of a mile or so emerged onto a rolling plateau of grassland stretching far to the north.

Salisbury Plain has an atmosphere all of its own, and one that changes minute by minute. History seems closer to the surface on the Plain, less obscured by recent change than almost anywhere else in Britain. Under grey skies it can be bleak, windswept and lonely – and so it was that first morning. Aside from the track I was on, there was hardly a sign that another human had ever come by. Indeed, the view can have changed little since the land was first cleared of trees some 5,000 years ago. I pulled over, took out my net, and began to walk. The Plain is slightly elevated above the surrounding landscape, so that the distant horizon seems to drop away, enhancing the impression that you are in an other-worldly place, raised above the crowded hubbub of everyday life. Rolling, empty grassland stretched in all directions, broken only by the occasional small clump of scrubby hawthorns or windswept, distorted beech trees. It was cool and windy, and as a result there were very few bees about, but the flowers were extraordinary. There were great swathes of familiar meadow and downland flowers such as red and white clover, lady's bedstraw, hawkbits, bird's-foot trefoil, ox-eye daisy, rock-rose, tormentil and salad burnet, but also a bewildering variety of less familiar species, some that I had never encountered before. Sainfoin grew in abundance, its delicate pink flower spikes swaying in the breeze. This member of the pea family was once widely grown as a fodder crop, but has fallen out of favour with farmers who no longer need nitrogen-fixing legumes in their crop rotation; to this day, Salisbury Plain is the only place in the UK I have seen sainfoin growing wild. There were many other unusual legumes – kidney vetch, dyer's greenweed, horseshoe vetch. The sward was dotted with ant mounds, Dalek-shaped hummocks veiled in purple thyme. In the disturbed areas near the tracks were huge patches of red bartsia, a scraggy little plant with nondescript purplish flowers, but much loved by bees. Tall blue spikes of viper's bugloss and the heavily-scented yellow spires of wild mignonette also lined the dusty tracks. Betony and black horehound sprouted from around the occasional patches of hawthorn and blackthorn scrub. This was bee paradise – and it seemed to go on for ever.

In fairness, as I explored I was soon to find that not all of the Plain is so rich in flowers. There are patches of arable land, some areas that have been 'improved' by adding fertilisers, some where the scrub has encroached and there is little but hawthorn. But the overall effect is a mosaic of flower-rich patches, some of them huge, within which it is always going to be easy for bees, butterflies and hoverflies to find some of their favourite flowers on which to feed. The Plain is the most flower-rich area I have ever explored in Britain – only parts of the machair in the Outer Hebrides give it a run for its money in terms of the density of wildflowers, but they are quite a bit smaller in extent.

As I walked, the sun began to break through, and the gusty wind dropped a little. A skylark began to sing high overhead. In moments the atmosphere changed from brooding and desolate to peaceful, enchantingly unspoiled, the warmth raising the mingled scents of a myriad herbs from the rabbit-cropped turf. The first bumblebee appeared, a worker white-tailed bumblebee visiting the sainfoin for nectar, and then as I dropped into a slightly sheltered fold in the land it suddenly seemed that there were bees everywhere, the swaying flowers humming with activity.

I had hoped to be falling over rare species, but initially all I could find were species I could see in my back garden in Southampton – lots of red-tails, white-tails, garden bumblebees and common carders. I caught the common carders for close inspection, as records showed that there should be both moss carders and brown-banded carders on the Plain, rare species that I had never seen, but all three carder bee species are very similar. According to the books, all are a rusty brown, but the common carder has black hairs on the sides of the abdomen, which the other two do not. The brown-banded carder should have a sprinkle of black hairs amongst the brown near the bases of its wings, and a distinct darker-brown band across the abdomen. The moss carder has no black hairs on its back or sides, and has a neat and tidy coat, giving it a velvety appearance or, according to some sources, 'teddy-bear good looks'. My apologies for bothering you with such tedious distinctions – I'm afraid that the life of an entomologist involves quite a lot of staring at tiny, seemingly inconsequential features, and sometimes also trying to make sense of subjective statements such as the degree of similarity to a cuddly toy. Don't all bumblebees look like cuddly toys? Anyway, I had studied all the books, particularly Frederick Sladen's *The Humble-Bee*, published in 1912 and still arguably the best book yet written on bumblebees, in which he lovingly describes their lives and habits and the minutiae of the differences between the British species. I had familiarised myself with these distinctions, and it didn't sound so difficult on paper, but in the field, with a buzzing bumblebee in a small glass pot, I found it extremely hard to see whether or where there were black hairs, even with a hand lens (I was eventually to realise that the process is made somewhat easier by stuffing tissue paper into the pot until the bee is gently squashed against the glass wall and can no longer buzz about). I spent the next couple of hours staring at a succession of brown bees, reluctantly concluding that all were just boring old common carders. It is an unfortunate and inconsiderate feature of bumblebees that the rare species mostly look like one of the common species, almost as if they want to play hard to find.

I was becoming a little despondent at the lack of rare bees and had circled back to the car with a view to trying my luck elsewhere when I spotted a bee that didn't look quite right – it was black with a red tail, superficially similar to a common red-tailed bumblebee, but the red was a little less red, the black a little less black, and its bottom a little more pointy than is usually the case. It was also a little too small for a red-tailed queen, but a bit too big for a worker. It was visiting some tatty red dead-nettles that I had almost parked on top of, and something about the way it flew looked different too. I snaffled it in the net, and potted it up for a closer look. Of course it buzzed about unhelpfully for a while but eventually it tired and sat still, and I was able to get a proper look at it. As soon as I got a good look at its legs, I knew what it was – a queen red-shanked carder bee, a cousin of the various brown carder bees I had been looking for. The stiff fringe of hairs that make the pollen baskets on the hind legs are orange in the red-shanked carder, but black in the red-tailed bumblebee – very distinctive at close range. I apologise once again for giving you all the morphological details, which may be detracting from my effectively conveying what an exciting moment this was – my first rare bumblebee, a BAP[4] species. Red-shanked carders had once been widespread in the south-east – old distribution maps showed them all over Hampshire amongst other southern counties, and Sladen described them as common in Kent – but I had never seen one.

By the time I had finished admiring and photographing her and had let her go, it had clouded over once again, and a gentle drizzle began to blow in from the west. I hung around for a while, getting steadily damper, and in the end decided to call it a day. But my appetite had been whetted. If I was ever going to see more of our rare bumblebee species, and find out more about them, then surely this was the place.

I wanted to know why some bumblebees had become so rare in Britain, while others remain widespread. It seemed to me that if I could understand more about the needs of these rare species

---

[4] These are species that are formally recognised as endangered – BAP stands for Biodiversity Action Plan, for these are species for which the government is supposed to have developed and implemented a conservation strategy. There were seven BAP bumblebees in the UK, but the scheme was discontinued in 2010.

and what had driven their declines, I would stand a better chance of coming up with solutions, so that we could prevent them continuing to decline, or maybe even bring them back in places where they have disappeared. I devised a plan for a summer of fieldwork that would, I hoped, enable me to discover at least the basics of the ecological needs of these bees. My idea was to conduct timed, one-hour searches for bees at as many sites on the Plain as I could find time for through the summer months. I would count and identify every bee I saw, what flower it was on, and whether it was collecting pollen, nectar or both. I would also count the abundance of all the different flower types at each site. The idea was to build up a picture of the relative distribution and abundance of both the common and the rare species, and to gather data on which flowers the different bees preferred to visit. I hoped that the Plain might offer a window on the past, a view of what Britain was like a hundred years ago when much of the countryside was covered in flowers and when these rare species were still quite common. For example, I might discover that red-shanked carder bees are particularly keen on pollen from kidney vetch or the nectar of betony, flowers that are no longer common in most of the UK. If so, I would have a simple explanation for their decline and a solution – plant more kidney vetch and betony. Such flowers could be included in flower-strips planted on farms as part of agri-environment schemes, and red-shanked carder bees might once more spread into the wider countryside and become common. Of course, life is rarely as simple or easy as this, but that was the basic idea. What is more, it made a fantastic excuse for spending a lot of the summer romping about the Plain with a butterfly net while being able to genuinely claim that I was working.

Over the next two months I was to encounter most of Britain's rarest bumblebees, as well as managing to conduct a thorough lunchtime survey of the pies available in the many quaint timbered pubs to be found in the pretty valleys that intersect the Plain. I did eventually find both brown-banded and moss carders, after many more hours of staring furiously at equally furious bees squashed between glass and tissue paper. The brown-banded turned out to be quite frequent in a few places, and with practice the brown band is fairly easy to spot and minimal squashing is required. Sadly, I only ever saw a handful of those teddy-bear moss carders, but at least they were there. I also found ruderal bumblebees, though once again it took a lot of agonising over subtle differences in their colour pattern and the shape of their head to be sure I wasn't looking at their much more common sister species, the garden bumblebee. I encountered a range of cuckoo bumblebees, five of the six known British species, beasts that employ the underhand tactic of invading the nests of other bumblebees, killing the queen and enslaving her poor workers as their own. I found a bumblebee species I had not expected to see – the broken-banded bumblebee, possibly the most unhelpfully named of bees, for it does not usually have the broken band after which it is named, whilst the very similar and common buff-tailed bumblebee often does have a broken band, especially when balding with age. With practice, the two can be separated by other subtle characteristics – for example the broken-banded has a reddish fringe to its white tail, while the buff-tail, which in the worker is actually white tailed, usually has a hint of a brownish fringe. (Sorry, I'm at it again – you'll appreciate by now why few people ever get fully to grips with identifying bumblebees.) The broken-banded bumblebee is mainly found in the hills of north and western Britain, which was why I had not been expecting it, but it turns out that there is an odd outlier population on Salisbury Plain – odd because the habitat could not be more different. This species was never given BAP status, but there is a good argument to be made that it should have been since it has declined enormously in the last fifty years, with the only reasonably strong population south of mid-Scotland now on the Plain.

In my exploration of the Plain I visited Imber, today a sombre, lifeless place. The church has been maintained and an annual service is held there, but otherwise the buildings have been slowly allowed to crumble. The old cob and timber-framed houses have rotted and collapsed, being absorbed back into the Plain. The brick-built houses have had their thatched roofs replaced with rusty corrugated tin to keep them standing so that they can be used in mock street battles, and the black sockets of their broken windows stare mournfully at what was once the village high street. It is hard to imagine the bustling little community that once lived there.

In early August the military operations shut down for two weeks, and I used the opportunity to sneak into the impact zone near the centre of the Plain. Artillery batteries several miles away in Larkhill pound the hell out of this site on a regular basis, producing a lunar landscape of craters in a central area just a few hundred metres across – reassuringly suggesting that they are usually good shots. The craters were sprouting arable weeds such as charlock and poppies – plants normally associated with the disturbance of ploughing, here thriving under the extreme disturbance of regular explosions. Surrounding this pockmarked area was dense scrub – for obvious reasons, domestic livestock such as sheep are not brought out here. Amongst the hawthorns numerous badger setts were evident – presumably the badgers occasionally get blown to smithereens, and perhaps some are a little hard of hearing from the terrible din they must regularly endure, but clearly this was not enough to deter them from living here. Without grazing animals this part of the Plain is slowly reverting to the woodland it would once have been 6,000 years ago. It was fascinating but slightly unnerving to visit a place that is almost never visited by humans – perhaps the only such place in Britain. One might argue that the scrub should be cleared, since it is displacing the many flowers and insects that thrive on the open downs. On the other hand, it is nice to see nature taking its own course for a change.

My most exciting bee find did not take place until late August, when I had all but given up hope of seeing this particular species. The shrill carder, named after its distinctively high-pitched buzz in flight, is arguably Britain's most endangered bumblebee. It was once found all over the south and east of Britain, but it declined rapidly as our flower-rich grasslands were destroyed, and it is now found at just a handful of sites, of which Salisbury Plain was reputed to be one. Unlike the other carders, the shrill carder has the decency to have a distinctive colour pattern – it is largely greyish brown, with a black stripe across the thorax and a reddish bottom. This may not sound wildly exciting, but at least it is different, and I was pretty sure I would know one if I saw it. As the summer wore on, and successive visits to the Plain failed to yield a single one, I began to wonder if they had gone extinct there. Then, one late afternoon as I was scouting amongst a scrappy patch of viper's bugloss near the eastern edge of the Plain, I found myself being circled by a small, high-pitched bee. I'd been struck by this odd behaviour on several occasions during my visits to the Plain – occasionally, and particularly when I was standing out in the open, bees would zoom around my head three or four times in a tight circle before careering off at speed. It felt a little bit as if I was being inspected, investigated as an interesting new feature of the landscape.[5] Anyway, on this occasion I flapped out wildly

---

[5] The following year I came back with most of my research team to investigate this in more detail. When a bee is circling one's head at speed it is hard to catch, but I had flapped at quite a few and been

with my net, and more by luck than judgement I caught the inquisitive bee and popped it into a pot – lo and behold, it was a rather tired-looking shrill carder worker. Before summer's end I saw two more, one more worker and a male – the latter very handsome, for the males have brighter colours than the workers.

In many ways, the best thing about exploring the Plain was not the bees but the other wildlife that I stumbled across. Outside of the tropics, I have rarely if ever seen so many butterflies. On my very first visit I saw plentiful marsh fritillaries, a very rare and declining species with checkerboard orange and black wings. Later in the summer things really got going with clouds of chalkhill blues, marbled whites, vivid sky-blue Adonis blues shimmering in the sunshine, beautifully camouflaged graylings, dark green fritillaries soaring on the breeze, little chocolate-coloured arguses zooming energetically amongst the flowers, and many more. The Plain is also famously rich in bird life, including two of Britain's oddest and rarest birds, although I only saw one of the two that summer. This was the stone curlew, an awkward-looking

---

struck by the fact that the bees I caught were almost invariably redtails and broken-banded bumblebees. Buff-tails, common carders and early bumblebees were much more common on the Plain than broken-banded bumblebees, yet none of the former ever seemed to indulge in this odd behaviour. With my team, we spent a couple of days in the summer of 2003 and 2004 doing nothing but standing in the middle of the Plain and attempting to catch the bees that circled our heads. My early impression was correct – some species seem much more prone to this behaviour than others. To this day we have little idea why they should differ – perhaps these species put more effort into memorising any new landmarks that they encounter, to aid their navigation. Or perhaps red-tails and broken-banded bumblebees are just more nosy.

chicken-sized creature with a disproportionately large head and bright yellow, slightly bulging eyes. I was lucky enough to spot a pair of them stalking around in a fallow field, but when I tried to creep close they saw me and gave a mournful, plaintive cry before galloping off on their long, gangly yellow legs. I didn't see the second species because it was not there in 2002, but it has since returned, the subject of an ongoing reintroduction programme – the great bustard. Bustards are odd creatures, in appearance a little reminiscent of grouse or turkeys, but in truth more closely related to cranes. The great bustard is the world's heaviest flying bird, with the males weighing up to twenty kilograms and standing over one metre tall. It is also perhaps the most ridiculous. To attract a mate, the males engage in a remarkable and bizarre display, in which they simultaneously inflate an air sac in their neck so that it becomes hugely distended, point their whiskery chin feathers to the sky, flip their wings upside-down and fold their large white tail forwards towards their head. The wildlife TV presenter Chris Packham once memorably described them as resembling a vicar in a tutu. They may hold this uncomfortable pose for many minutes, occasionally shivering their plumage for added drama. As if this doesn't make them look foolish enough, their call sounds rather as if they are simultaneously sneezing and breaking wind – the combined effect of all of which is apparently appealing if you happen to be a female great bustard. Let's hope so, for the males' sakes.

The great bustard is a creature of wide, open spaces. Once found in Wiltshire and East Anglia, and abroad in the Steppes of Russia and the great plains of Eastern Europe and Spain, its large size made it an irresistible target for hunters, and so it has been exterminated in many areas. In a crowded country such as Britain it stood little chance, particularly in the great nineteenth-century age of hunting, and the last one was shot in 1832. The hunters must surely have known that the great bustards were disappearing, but presumably they just couldn't help themselves. One hundred and seventy-two years later, in 2004, a team of volunteers calling themselves the Great Bustard Group began a reintroduction programme, using young birds from Russia. The supply of birds was very limited, since they were only allowed to rescue eggs from nests that were in arable fields and were about to be destroyed by harvesting of the surrounding crop. The birds, when hatched, then had to be reared in captivity. Each year about twenty young birds were released, and by 2009 some of the released birds had survived to adulthood and produced the first wild chicks to hatch in Britain for nearly 180 years, although sadly they didn't make it through the following winter. Adult survival has not been great either – the young birds suffer from predation, and also they have a habit of flying into fences. I visited the release area a few years ago and one of the staff explained that such heavy birds must take off and land at a shallow angle, and that while doing so they are prone to collisions with wire fences, which presumably they do not see until it is too late. The good news was that there seemed to be plenty of suitable food for them, for all of the corpses that had been retrieved along fence lines were in excellent health – aside from being dead.

The reintroduction continues to the present, and like that of the reintroduction of the short-haired bumblebee to Kent – a project with which I have been personally involved – the eventual outcome remains uncertain. They have recently switched to Spain for the source population for the release, following genetic studies that showed the Spanish population was most similar to that which had once lived in the UK. There is also a suspicion that the Russian birds have a strong instinct to migrate south in winter, causing some birds to leave the site and then find themselves in unsuitable areas where they stand little chance of survival. A mix of Spanish and Russian birds now exists on the Plain – I wonder how a Spanish female might react to the exotic sound of a Russian male's flatulent sneeze, and vice versa. Unfortunately, to my knowledge no home-grown chicks have yet survived a winter – hopefully it will happen eventually. A recent Tweet described seeing a drove (the collective noun for bustards) of five males flying past Stonehenge – what a magnificent sight that must have been. Perhaps one day a viable population will establish itself, though I suspect it will always be small and teetering on the edge of extinction, unless bustards can somehow master the skill of vertical take-off and landing.

My favourite non-bee denizens of the Plain are rather smaller than the bustards, and have a peculiar mutualistic relationship with the tanks. I had been told to look out for these creatures, and so on my visits to the Plain that first summer I peered excitedly into every muddy puddle I came across. As you might imagine, the tanks churn up the tracks across the Plain whenever it rains, taking huge gouges out of the earth which then fill up with rainwater. By late spring these tend to be drying up, pea-soup green and stagnant, and this is the perfect time to see fairy shrimps. It was on just my second visit to the Plain that I saw my first one – a semi-transparent, greenish-brown creature perhaps three or four centimetres long, lying on its back, and using its innumerable bristly legs to rhythmically waft its way along just beneath the surface of the water. The eyes were black and protruding on stalks from each side of the head, and the tail long and forked – altogether a most bizarre creature. I got a little too close and it darted down into the murky depths where I could not see it, but after a few minutes it rowed itself back into sight. They turned out to be quite common on the Plain, though I never tired of looking for them. Sadly, this is one of the only places in Britain where they can reliably be found – the filling in of temporary ponds and the pollution of those that remain mean that the number of surviving populations of this species has dwindled to about half a dozen in the south-west of Britain. But the tank activity on the Plain provides them with plentiful habitat, and also a means of transport. Fairy shrimps are highly specialised creatures that can survive only in temporary ponds – they are virtually defenceless against predators such as fish or dragonfly larvae, so they can only survive in ponds that exist for such short periods that predators do not have time to arrive. Their eggs hatch in the winter rains as the puddles fill, and the shrimps grow fast in spring, using their many legs to sieve the water for the algae, bacteria and so on which are their food. Before their pond dries out in summer they must complete their development and produce eggs, which survive being dried out and then sit in the summer dust of the Plain waiting for the water to return. The eggs can survive for many years if the conditions aren't right for hatching. In the past the eggs, mixed up in the sticky mud of a drying summer puddle, may have been spread from place to place on the legs or bodies of large mammals such as aurochs or boar – the latter's constant truffling about in the earth may also have helped create small ponds for them to live in. More recently, I'd imagine that horses and carts passing along unpaved roads would have created a network of suitable habitat all over Britain, which the advent of tarmac did away with. These days, on Salisbury Plain at least, the mud-spattered tanks and other military vehicles are thought to do the job pretty effectively, spreading eggs along the tracks so that a high proportion of the suitable puddles are occupied. This serendipitous synergy between army vehicles and a tiny, delicate shrimp epitomises the relationship between the army and wildlife on the Plain. The army purchasing strategy in 1897 didn't set out to create a huge nature reserve – one imagines that this would have been somewhere near the last thing on their minds – but that is nonetheless what happened. Although the purpose of their estate remains primarily that of providing a training ground for the troops, these days the army is sensitive to the needs of wildlife, and it has adapted to a new role as custodian of the rare plants and animals that flourish on the Plain. Some of the army personnel are wildlife enthusiasts, spending their spare time mapping and recording the populations of butterflies and flowers, in between dashing about in tanks and practising shooting one another.

To return to the bees, you may be wondering what exactly I found out from my extended jolly on the Plain. You'll recall that my aim was to find out more about the requirements of our rare bumblebees, particularly with regard to their preferred flowers, so that we might better look after them. I did end up with a very large spreadsheet, with the many different flowers listed down the side, the bee species along the top, and an awful lot of numbers in the middle, indicating how many of each bee had visited each flower type. Of course, most of the records I amassed were for common bumblebees – red-tails, in particular, which are inordinately abundant on the Plain. The large majority of all the bees I recorded were visiting quite a small number of plants: viper's bugloss, red clover, sainfoin, melilot, knapweeds, red bartsia and thyme. Frustratingly, data for the rare species were sparse – even on the Plain, the largest tract of flower-rich grassland in Western Europe, the rare bumblebees were still scarce. For the shrill carder, my entire data set included one male on black horehound, one worker on red bartsia, and one worker flying round my head – not a very comprehensive picture of the foraging behaviour of the species, and certainly no basis for recommending the mass planting of bartsia and horehound all over the UK, though I can think of worse things to do. Likewise, I had just two sightings of ruderal bumblebees, three of red-shanked carders and four of moss carders – a depressingly small haul from a whole summer of work.

However, I did have slightly more respectable data sets for some other rare species, enough to suggest that brown-banded carders seem to be legume specialists, visiting mainly red clover, sainfoin, melilot and bird's-foot trefoil, and that the broken-banded bumble-bees were mostly on sainfoin, red bartsia and field scabious.

On my last day on the Plain I reviewed what I had found over the summer, going through my messily scribbled notebook while tucking into a lovely dish of locally caught potted pike (man cannot live on pie alone). All in all, it had been an interesting start, and it had been wonderful to at least get to see these rarities on the wing, but it was also clear that I needed to go further afield if I really wanted to find out more about these elusive beasts. I needed to find somewhere where they were still as common as in Sladen's day, which was clearly not going to be easy.

CHAPTER TWO

Benbecula and the Great
Yellow Bumblebee

On earth there is no heaven, but there are pieces of it.

Jules Renard

As our tiny, noisy, twin-prop eighteen-seater plane banked hard right to come in for the runway at Benbecula airport, I looked directly down on a scene straight out of a Caribbean holiday brochure: a crescent of dazzling white sand lapped by a crystal-clear turquoise sea. Above the beach the dunes were splashed with colour: drifts of yellow, purple and red flowers. No palm trees to be sure, but in every other respect one might have been arriving on a subtropical isle, ready for a week of relaxation, sunshine, piña coladas and reggae music. I was sharply returned to reality when, a few minutes later, I stepped out of the plane into the icy wind; this was most definitely not Antigua.

Benbecula airport is tiny, but it does at least have a runway. If you fly to the nearby island of Barra, your plane lands on the beach itself, and flights can only come in when the tide is out. This is the most remote corner of the British Isles, the Outer Hebrides, a chain of low-lying islands sixty kilometres to the west of the Scottish mainland in the North Atlantic. It was August 2005, and I was here on a mission: to see the great yellow bumble-bee, properly known as *Bombus distinguendus*, a contender with the shrill carder for the title of Britain's rarest surviving bumblebee.

One hundred years ago, great yellows were found all over the UK. They were never particularly abundant in the south, but there are records from almost every county from Cornwall to Sutherland, and Norfolk to Pembrokeshire. They were always more common in the north, seemingly preferring cool, wet parts of the country, though not the uplands. Sadly, the great yellow did not much enjoy the twentieth century. Populations in the south started to disappear from about 1940 onwards. Even the flowery ranges of Salisbury Plain were not enough to support them, and within forty years the species had disappeared from the entirety of England and Wales and from most of mainland Scotland. By the end of the millennium they were only to be found on a few Hebridean Islands, Orkney and in a scattering of tiny populations along the far north coast of Caithness and Sutherland. Now, the strongest surviving populations are in the Uists, the central part of the Outer Hebridean chain, comprising North and South Uist with Benbecula sandwiched between the two. Which was, of course, why I had chosen to come here: to see the last of the great yellows, to learn more about this scarcest of British bumblebees, and perhaps to do something to help save them from extinction.

I was met at the airport by my PhD student Ben Darvill, who had spent the previous two summers in the Hebrides studying the bumblebees to be found there and taking DNA samples for genetic work. He was trying to find out whether these isolated island populations were suffering from inbreeding, and the gradual loss of genetic diversity that can occur when a population is small and cut off from gene flow – isolated from others of its kind. Ben was in his decrepit VW camper van, a vehicle so ancient and damp that moss and occasionally small fungi sprouted from the insides of the windowsills. It was to be our accommodation for the next week.

I had expected it to be something of a struggle to find great yellow bumblebees. Seeing most scarce, highly endangered species requires considerable dedication – it had taken me nearly two months to see my first shrill carder on Salisbury Plain – but in the Uists great yellow bumblebees were everywhere that year. It was somehow almost disappointing. I saw my first one thirty yards from the airport, on the large pink flowers of a hedge of bushy roses growing along the edge of a garden. She was a worker, scampering around amongst the mass of stamens, emitting high-pitched buzzes to shake free the pollen. Tremendously excited, I spent a long time photographing her from every conceivable angle, anxious that she might be the only one I saw on my trip. However, I had to admit that she didn't really live up to the name – she wasn't particularly big, and nor was she especially yellow. The 'average-sized, straw-coloured bumblebee' would be a more appropriate name, if less catchy. She was rather pretty, nonetheless: tawny-blond all over apart from a clear black band across the middle of the thorax, the great yellow is one of our easiest bumblebees to identify.

We headed south from Benbecula, away from the ugly gaggle of concrete buildings near the airport, and I was struck once again by the extraordinary beauty of the area. To our right we passed a succession of perfect, white-sand beaches, with not a single soul on any one of them. The sun sparkled off the sea beyond. To our left, in the distance, gentle hills rose, cloaked in purple heather. The road itself passed through a habitat known as machair, the secret of the survival of great yellow bumblebees in the Hebrides. Machair is one of the rarest habitats in the world – in fact, nearly all of it that exists anywhere is in the west of Scotland and the west of Ireland. It is formed from a flat plain of wind-blown shell sand – tiny grains of broken shells that have been ground up by the action of the oceans over millennia. The machair plain is sheltered from the prevailing westerlies by the dunes above the beach, but catches fine grains of sand that settle in their lee. The shell sand is inherently alkaline, and very poor at retaining nutrients, which wash through and leach away in the frequent rain. Further inland still, the bedrock rises into low, ancient hills that are clothed in a thin layer of acidic peat, which supports mosses and heathers, the more familiar moorland vegetation found across much of Scotland. So the machair is a flat ribbon of land, never more than a kilometre or two wide, running down the west coast of these islands, sandwiched between the dunes and the sea on the west and the hills on the east.

A flat plain of nutrient-poor sand may not sound terribly promising, but this plain supports a wealth of wildflowers, and hence a diverse and thriving bee community, along with much other wildlife. Counter-intuitively, wildflowers proliferate where nutrients are scarce. In particular, legumes dominate – their root nodules, packed with nitrogen-fixing rhizobium bacteria, enable them to take nitrogen from the air and turn it into useable nitrates, vital for building proteins. Most other plants such as grasses cannot do this, and so they grow slowly on the machair, leaving lots of space and light for the legumes to flourish. As a result, as we drove along, the landscape on both sides was a carpet of red clover, white clover, tufted vetch, kidney vetch, and bird's-foot trefoil – all plants that are also abundant on Salisbury Plain, and all of them well loved by bumblebees, with different bee species tending to have slightly different favourites. If bumblebees go to heaven, it might well look something like this.

We pulled over and strolled through the knee-high flowers, butterfly nets and cameras at the ready. Ben had endured weeks of rain before I arrived, sheltering in his mouldy van: there isn't a lot to do in Benbecula in the rain. I was incredibly lucky, for the weather while I was there was atypically fine; the sun did its best to blaze down from a cloudless sky, though it never got terribly warm, for at this latitude the sun never gets too high in the sky. A short-eared owl glided by, hunting for voles. It was odd to see an owl hunting in the daytime, but that is normal for this species. Perhaps at this latitude the summer nights are so short that they have had to become active in the day. As we walked, our feet disturbed hundreds of bumblebees, busy extracting pollen and nectar, making hay while the sun shone. I was particularly excited to see moss carder bees in abundance. In mainland Britain they are very hard to find, and they are also exceedingly tricky to distinguish in the field from the common carder and the brown-banded carder, as I had discovered on the Plain. In the Outer Hebrides, there is no need to worry about subtle distinctions, for the other carder species do not live out here, and in any case Outer Hebridean moss carders look quite different to their mainland counterparts. We have no idea why they should be different, but here they have a rich chestnut-coloured thorax and black underbelly, which combined with their 'teddy-bear good looks' makes them very handsome bees indeed. We also have no explanation as to why, on the most outer Hebridean islands, moss carder bees are by far the most common bees, bucking the trend found elsewhere across Europe. Here, they seem to be the dominant species, the superior competitor, while elsewhere in the UK they are invariably rare and seem to be hanging on by the skin of their mandibles.

Moss carders are not the only bumblebee to have a unique colour form only found in the Hebrides. We also saw heath bumble-bees, the queens of which are normally rather small with three yellow stripes and a white tail. Here, the queens were gigantic, with reddish brown tails. I was initially quite confused as they superficially looked like the big buff-tailed queens one commonly sees in most of Britain, but I knew that buff-tails don't occur in the Hebrides.

We spent the whole afternoon in that sea of flowers – I spent much of my time on my belly, camera at the ready, stalking great yellows and moss carders, and trying not to squash too many flowers. They were mostly workers, but there were still a few queens about, even in august, for the season up here starts very late, with most queens remaining in hibernation until June. The queen great yellows were still not really big enough to justify their name, but were nonetheless extremely splendid insects. More by luck than judgement, some of the photos that I took on that day came out unusually well, and they have been used many times since in promotional literature by the Bumblebee Conservation Trust and on the cover of my scientific tome on bumblebees. Every time I see them, I am reminded of lying belly down amongst the flowers, the Hebridean sun on my back and bees buzzing all around.

We spent the next few days touring to the bottom of South Uist and back up to the top of North Uist, recording what we saw, identifying the flowers that the different bees were feeding on to add to my growing database, taking genetic samples,[6] and snapping as many photographs as we could. I was struck by how similar the flora was to that of Salisbury Plain – most of the common flowers were the same, despite the big difference in climate and geology. In fact, I should not have been surprised – both habitats are nutrient poor and well drained, and both are growing on the calcareous shells of long-dead sea creatures. Were it not for the distant roar of the surf, and the consistently chilly air

---

[6] To do this we chop the last segment off one of the legs of the bee – it seems a little harsh, but the poor bees seem more or less unaffected.

temperature, one could easily have been on Salisbury Plain or one of the unspoiled fragments of flower-rich grassland on the South Downs. In addition to the many legumes, there was lots of knapweed, the tall purple flowers favoured by male bumblebees, perhaps because their large flowers on sturdy stems make perfect platforms on which groups of males can lounge about drinking nectar before going in search of a mate. There was also plentiful yellow rattle, a 'hemi-parasite' that parasitises grasses and sucks nutrients from their roots.[7] There were also lots of arable weeds such as corn marigold, poppies and corncockle. Arable weeds have had a tough time in recent decades, for these are annual species that used to thrive in the disturbance created by annual ploughing and cropping. In days gone by, cereal fields were routinely washed with the blue of cornflowers, or red with swathes of poppies. Grains collected for sowing the next year were often contaminated with wildflower seed, so they were accidentally spread from place to place by man. These weeds would also thrive when fields were left fallow for a year, flowering in profusion and scattering seeds that could then sit in the soil for several years waiting for a chance to germinate. Poppy seeds are particularly good at this, being able to sit, inactive, in the soil for decades until the right moment arrives. Modern seed-cleaning methods now remove unwanted wildflower seeds, and most cereal crops are sprayed with herbicides that kill all broad-leaved plants, so most arable weeds have declined enormously. Some, such as downy hemp nettle and thorowax, are now extinct in the UK, while many others are declining and some teeter on the edge: corncockle, shepherd's-needle, Venus's looking-glass, pheasant's eye,

---

[7] In *A Buzz in the Meadow*, I describe how this plant may be used to help recreate flower-rich grasslands, and also how, in the Alps, specialist nectar-robbing bumblebees learn from each other how to steal nectar from rattle.

and so on. Surely they are worth saving for the evocative names alone?

The reason that arable weeds remain common in the machair is due to the low-intensity farming practices that still take place there. Farms in the region are called crofts, and are often tiny, a few acres of fenced fields close to the farmhouse, which is traditionally a low stone hut with enormously thick walls and often just one or two rooms inside, although most such buildings have now been replaced with more modern dwellings. The crofters usually share ownership and management of more extensive areas of machair and moorland, traditionally grazing the machair in winter and putting their sheep up on to the hills in summer. Patches of machair are occasionally planted with crops, small patches of potatoes, oats and rye, traditionally fertilised with seaweed carried up from the strand line on the beach, with these patches commonly being left fallow for a year or two after cropping. Because of their remoteness and the small scale of the farming, the use of artificial fertilisers and pesticides caught on more slowly out here, and remains much lower than on the mainland. Seen from above, the plain of the machair is chequered with small patches of crops interspersed with patches of fallow land of varying ages, set in a matrix of clover-rich pasture. It provides an increasingly rare example of how man's activities can promote wildlife, creating a wonderful mosaic of habitats that collectively support a great diversity of life.

Looking at a distribution map for the great yellow bumblebee today, one could be excused for concluding that this is clearly a coastal species, for there are no populations more than about eight kilometres from the sea, most much less than that. One might even think it a species that specialises in living on the machair, for this is certainly where it is most common. However, a moment's reflection reveals that this is clearly not true – old records suggest that the great yellow used to be found scattered across England; for example, there are records from Warwickshire, a county that is not known for its extensive coastline or its machair grasslands. In Europe, the great yellow can be found in southern Poland, more than 1,000 kilometres from the nearest coast. It is not that this species prefers the coast, it is more that the flowers on which it depends are now only found in sufficient abundance at coastal sites, at least in the north of the UK where this species seems to be most at home.

It is not really clear why the great yellow is so rare. Watching them in the Uists, it quickly becomes obvious that they do not have particularly unusual flower requirements. They have relatively long tongues for bumblebees, and so feed on deep flowers such as yellow rattle, kidney vetch, tufted vetch, red clovers and knapweeds, but they will also opportunistically visit a range of other flowers, including the domesticated roses I saw them on in gardens. Neither do they seem to be especially fussy about where they nest – although their nests are not found often, those few records we have (some of them thanks to a sniffer dog that we once trained for the purpose) suggest that they often use old rabbit burrows, something that is not in short supply in most of the UK. Of course, the flower-rich grasslands that the great yellow prefers are much less common than they used to be a hundred years ago, when the UK was awash with hay-meadows and chalk downland, but we do not yet have a convincing explanation as to why the great yellow has been hit harder than almost any other bumblebee species. One long-tongued bumblebee species, the garden bumblebee, remains fairly common all over the UK. According to the thousands of records we have amassed over the years, it feeds on a near-identical range of flowers to the great yellow. Clearly flower preferences alone do not explain why some species are more sensitive to habitat loss than others.

Sadly (for the wildlife, at least), crofting is changing. Most crofters are over sixty years old, and they are not being replaced, as their children tend to choose not to follow in their footsteps. Crofting is a hard life, for a traditional small croft cannot support a family – most crofters also have part-time jobs to supplement their income, and even then they will always live very modest lives. These tiny farms qualify for only small farming subsidies, while much larger sums are dished out to relatively wealthy farmers elsewhere in Europe. This is not a glamorous career choice. Children growing up today watch the television or surf the Internet, and realise that there is more to life than eking out a living on the edge of the world. They want pubs and clubs, shops, excitement: and who can blame them? So they leave, heading off to the bright lights of Glasgow or London, and slowly the crofts are falling into disuse. Abandonment of cropping leads to the loss of the arable weeds, and cessation of grazing by sheep in the winter leads to the machair vegetation becoming tall, dominated by coarse rushes, and developing a thick thatch of dead grass and other vegetation, so that floral diversity slowly declines.

While some crofts have been abandoned, others have been acquired and merged into larger operations, often dependent primarily on sheep ranching. The tradition of moving sheep to the hills in summer is often no longer practised as it is time consuming and more difficult to keep an eye on the animals, so many of the larger ranching operations maintain high densities of sheep on the machair right through the summer. This is pretty hopeless for bees as the sheep keep the sward grazed down close to the ground. Almost nothing can flower since sheep love to consume buds, so there is no food for pollinating insects.

These shifts in farming practices have had similar negative impacts on another iconic Hebridean creature, the corncrake, a species with a history of declines in the UK that closely parallels that of the great yellow. This odd bird, related to the moorhen, prefers to nest in long grass: cereal fields, hay-fields and ungrazed summer machair. It is not an especially remarkable creature to look at: bantam-sized, and beautifully camouflaged in fawns and russet with darker flecks. It used to be common across Britain, but declined swiftly as new crop varieties and availability of fertilisers enabled crops to be harvested earlier. Just as with the great bustard, which also likes to nest in open fields, there was no longer time for the chicks to fledge before the crop was harvested. Thousands of nests were simply mown over, the eggs and chicks destroyed. The switch from hay, which is cut in summer, to silage,[8] which can be cut several times through spring and summer, was particularly devastating. In Northern Ireland, a campaign to encourage farmers to produce silage instead of hay, and to encourage them to keep more sheep, resulted in an 80 per cent drop in the corncrake population in just three years between 1988 and 1991. The Hebrides is now the last UK stronghold for this species, though only a few hundred remain. It is a shy bird, rarely leaving the long grass, so we did not see one on that trip, but I have been lucky enough to hear them and briefly see them since then on a spring visit to the Isle of Oronsay in the Inner Hebrides, an island that is entirely managed by the Royal Society for the Protection of Birds to encourage rare birds such as chough and corncrakes.

---

[8] Silage is made by cutting fresh grass and packing it densely in clamps or round bales, sealed with a plastic sheet. In these anaerobic conditions the grass keeps well while gently fermenting, producing a smelly brown mulch that cows and sheep readily consume. Unlike hay it does not require fine weather for drying, and with plentiful fertilisers to encourage the grass to grow several cuts can be taken per year, providing far more fodder per acre than a hay meadow. Good news for farmers, but not so good for bees, bustards or corncrakes.

In the breeding season in mid spring, just after the corncrakes have arrived back from Africa on migration, the males attempt to attract females with their call – a dry rattle that sounds more like the work of a giant grasshopper than that of a bird. Unlike the flamboyant great bustard, the males are shy, secretive creatures, calling from deep cover and mostly at night. On Oronsay their night-time calls were so incessant that several stiff measures of Laphroaig whisky were needed before I could get off to sleep.

There is no point in blaming anyone for the changes that led to the demise of the corncrake and the great yellow bumblebee. The world has moved on, and inevitably traditional ways of life are lost as new ones develop. The challenge for society is working out what to do. We could heavily subsidise traditional practices, essentially paying crofters to carry on doing what they used to do in the hope that this props up rural communities and helps to preserve the wildlife associated with crofting. This would be expensive for taxpayers, and runs the risk of creating a Disneyesque parody of rural life. On the other hand, if we allow crofting to disappear then some of the unique wildlife of the machair is likely to disappear. The RSPB now manage some big chunks of the Hebrides, including a sizeable area of machair on North Uist, and with their considerable resources, and working closely with locals, hopefully they can conserve some of the character of the land and the biodiversity it supports.

For the moment at least, great yellows and moss carders do seem to be holding their own in the Uists. However, there are other threats looming in their future. These populations are now highly isolated; where once there were populations on the Inner Hebrides and the mainland, sources of immigrants to bring in 'new blood' – genes from outside – now there are none. There is a distinct danger that these populations will slowly become inbred over time, losing vital genetic diversity, and that they will eventually fizzle out as a result, regardless of whether the habitat survives or not. This was exactly what Ben was studying, by taking DNA samples from each bumblebee species to be found on the different Hebridean islands, so that we could measure how much genetic diversity remains, and estimate how often bees move between the different islands.

We were particularly keen to get genetic samples from the most isolated bumblebee populations of all, on the Monach Isles. From North Uist the Monachs can be seen in the distance as a grey smudge on the western horizon. A cluster of low-lying islands, they lie about ten kilometres further west into the North Atlantic than the main chain of the Outer Hebrides. They are not currently inhabited, and hence there is no regular transport, so we were unclear as to how we might get out there.

As we travelled the Uists and Benbecula, collecting genetic samples and information on the basic ecology of the different bee species, we asked the few folk that we encountered whether they knew of a way out to the Monachs. For several days we drew a blank, though the conversations were worth it to hear the lovely, gentle accent of the locals – to my ear a blend of Highland Scottish and southern Irish. Many of these people were more comfortable with their native Gaelic than speaking English. Eventually we chanced to ask in the hardware store on Benbecula. The store's owner, a tall, weatherbeaten character, introduced himself as Ronald McDonald, forcing me to splutter as I choked back laughter – I guess perhaps he was used to the reaction. It would have been less funny had he not had a large red nose. Ronald happened to know a Donald MacDonald, no relation, who kept a flock of sheep on the Monachs, and went out to check on them periodically. He very kindly offered to look up Donald's number in the telephone directory. This took longer than you might think, for although the local Benbecula telephone directory is little more than a pamphlet compared to the sturdy doorstop that most of us are used to on the mainland, the MacDonald section covered several pages, and there were dozens of entries for 'MacDonald, D'. Eventually we found the correct number, but could then get no reply. It took a couple of days of repeated calls before we eventually got an answer, only for this to turn out to be a dead end; Donald had recently been out to tend his flock and was not planning to go back in the near future. We'd missed our chance.

We were running out of time, with only two days before my return flight, when, by chance, a casual conversation with some birdwatchers revealed that they had managed to charter a boat to go out to the Monachs the very next day. As luck would have it, there was room on board for two more, and they were very happy to split the fee. We arrived at our rendezvous point on a remote beach a little early the following morning, just after dawn, and as we waited we were lucky enough to see an otter come in from the sea and lollop up the beach not fifty metres away. It didn't seem to notice us, and stopped to wash itself for a moment before disappearing into the tussocky dunes. Shortly afterwards the birders arrived, chatting excitedly and laden with a plethora of binoculars and camera bags, and then our lift hove into view around the headland to the north, a large rubber inflatable with two huge Yamaha outboard motors, piloted by a capable-looking guy named Craig sensibly kitted out in a thick black rubber dry suit. We waded in to the chilly waters to climb aboard, and soon we were skimming across the choppy sea towards the distant Monachs. It felt like a fairground ride as we scudded at speed over the peaks of the waves, shivering in our damp clothes, eyes squinting against the stinging spray. Within twenty minutes we entered the smoother water in the lee of the islands and our pilot powered down the engines.

There is something magical about exploring an uninhabited island that brings out the inner child in all of us. We were eager to see what we could find, and we scattered in different directions, Ben and I with our butterfly nets at the ready, the others brandishing their binoculars and long-lensed cameras. It was a beautiful day, and I quickly warmed up as I headed to the highest spot I could see, an enormous system of dunes that rose perhaps forty metres above the beach. From the top, I could see across the entire island chain. The Monachs really consist of three separate islands which together stretch over about four kilometres, joined at low tide by bars of white shell-sand, plus numerous smaller rocky outcrops protruding from the surrounding sea. The islands once supported a small community of perhaps one hundred people, fishermen and crofters, all of them somehow eking out a living here, surviving on what they could grow in the thin soil or catch from the sea.[9] There was even a tiny school and a nunnery. Rectangles of crumbling stone walls are all that remain of their houses and livestock pens – the last people left in 1942. Until the fifteenth century, it is said that it was possible to walk from the Monachs to North Uist along sandbars at low tide, though it must have been a perilous journey rushing across the ten kilometres before the sea returned.

[9] The islands have even been used as an improvised prison. In 1732, a Scottish judge and politician named Lord Grange fell out with his wife, to whom he had been married for twenty-five years and who had borne him an impressive nine children. She accused him – probably correctly – of infidelity and treasonable plotting against the government, and to shut her up he had her kidnapped and taken to the Monachs. After two years he must have decided that even the Monachs were not remote enough, for he had her moved to St Kilda, another sixty kilometres further west, where the poor lady lived for another ten years.

Legend has it that this sandbar was swept away by a storm surge. There is no protected harbour on the islands, so from then on these folk must have been cut off from outside contact whenever the seas were rough, which out here is likely to have been much of the time.

Today there are no people, but the islands are infested with sheep. From the top of the dunes I could see hundreds of them, heads down, munching away. There seemed to be far more than such tiny islands could support. The flatter areas of the islands, inland from the dunes, are classified as machair, but this was a pretty poor sort of machair compared to that which we had seen on the Uists. It was grazed so intensively that the vegetation rose perhaps half a centimetre above the soil. The plants were miniaturised: bonsais, their leaves tiny, their stems prostrate against the ground. Normally, plants compete with one another, growing upwards to reach the light and shade their neighbours. Here, the plants pressed themselves against the ground, vying to get away from the endlessly nibbling lips of the sheep. There were quite a variety of plants; from their minuscule leaves I could identify red and white clovers, vetches and trefoils, but none were in flower. Any flower bud would be eaten long before it could open. In places, the vegetation had been stripped away completely, allowing the wind to blow out the soil, forming hollowed out bunkers.

I have a friend who disdainfully refers to sheep as woolly maggots. Environmentalist and writer George Monbiot has written at length about the adverse impacts that overgrazing by sheep (and deer) has had on the uplands of Britain, destroying the vegetation, preventing young trees from getting established, turning vast areas into monotonous grassy swards with almost no wildlife remaining, and compacting the soil surface so that rainwater flashes off the hills to cause floods many miles downstream. He argues eloquently that such farming employs only tiny numbers of people, contributes very little to the economy, and yet is heavily supported by subsidies of taxpayers' money. Why should we pay to support such a destructive practice? As I explored the Monachs, I couldn't help but think that he may have a point. These islands are classified as 'Sites of Special Scientific Interest' and are also a National Nature Reserve, and hence ought to be protected areas, managed for wildlife, where flowers, bees and all the other creatures associated with the machair could thrive in peace. This was what I had imagined they would be like. Instead, they were a barren, insect-free bowling green. It was disappointing, to say the least.

The only plants able to grow upwards and flower were spear thistles, protected by their fierce spines, and in places the thistles were taking over, forming dense stands that were extremely uncomfortable to push through. In fact, the thistles turned out to be a life-saver for us, for it was on their flowers that we did find a few bumblebees, though there were 'only' moss carders, mostly males. How they managed to survive through the year on such sparse resources was beyond me. Perhaps this was the tail-end of a dying population. We took our bee foot samples, and when Ben analysed them back in the lab it turned out that many of the males were diploids. Sex in bees is determined in a rather bizarre way, quite different to our own. In bees, males normally develop from unfertilised eggs, and so have half the normal amount of DNA, with just one copy of each chromosome; they are 'haploid', in scientific jargon. Females develop from fertilised eggs, and so have the full complement of DNA with two copies of every chromosome; they are 'diploid'. This all works very reliably in a large, healthy population with lots of genetic diversity, but it malfunctions in inbred populations. The reason for this is that sex is actually determined by a single gene, which occurs in the population as dozens or even hundreds of different forms, known as alleles. Any individual that has just one allele becomes male, and if it has two different copies it becomes female. Haploid individuals, with just one copy of each chromosome, must by default have only one allele of this gene, so they must be male. Usually diploids have two different alleles, since the chances of them getting two identical copies from their two different parents are slim; hence diploids are normally female. The problem arises when the population is small, for in small populations genetic diversity is lost; the number of different copies of any particular gene is steadily eroded, a process known as genetic drift. If the gene that determines sex loses its genetic diversity, so that there are only a few different alleles left in the population, then it becomes common for a queen bee to mate with a male who has the same allele as herself. When she lays fertilised eggs, intended to be daughter workers, half of them inherit identical copies of the gene and instead become diploid males. Since these don't do any work in the colony (in bumblebees, males never do any work to speak of, unless you count copulation), she is losing half her workforce, and her nest is likely to expire. The presence of so many diploid males on the Monachs was not a good sign for this remnant population of moss carders.

Yet despite the sheep and the depressing lack of all but a few inbred bees, I had an idyllic day. The tide was low at lunchtime and I made my way out across the sandbars to the second and then the third and endmost island. I ate my packed lunch, a rather low-quality but nonetheless enjoyable pepper steak pie, sitting on the dunes looking west towards Nova Scotia. On my way back I was vomited on by a young fulmar, not an experience I would choose to repeat but worth doing once just for the novelty. I'd been making my way along the north shore, between the dunes on my right and a rocky tidal ledge to my left, and had not noticed the large burrow amongst the tussocky marram grass until it was too late. I'd not heard a sound for hours apart from the lapping of the waves, the call of a few gulls overhead and the baaing of the sheep, so I nearly jumped out of my skin when there was a loud, guttural, retching bark close by and an instant later my leg was splattered with a hefty dollop of half-digested fish. The stench was horrendous. Apparently, surprising walkers in this way is a favourite pastime of fulmar chicks. The parents create nests in rabbit burrows close to the shore, sensibly preferring to nest on offshore islands where there are few predators. Once the chicks are old enough they are left alone during the day; fluffy, off-white little monsters, they sit in the entrance to their burrow waiting for passers-by, whom they target with surprisingly accurate jets of projectile vomit. Of course it is presumably meant to ward away predators; I was not inclined to go too near to a burrow again, and after being forced to smell their vomit I would certainly not fancy trying to eat one, so I guess it works pretty well. It took me ages to wash my shorts clean in the shallows, and despite my efforts the smell stuck with me for days (my apologies to the lady who had to sit next to me on the plane home). There were dozens more fulmar nests along that section of coast, and I had to pick my way between them very carefully.

A little while later as I climbed through the dunes, having left the fulmars behind, I became aware of a distant clamour; incongruously, it sounded like a primary-school playground in the lunch hour, a hubbub of noise that slowly swelled as I approached the top of a tall dune system. After the fulmar experience I was a little cautious, wondering what other noxious creatures might be lurking in wait for me, and so I started to creep slowly through the dunes, hunched down in what I recall from my experiences in our school's ludicrously incompetent cadet force as being called the 'monkey run' position. As the noise got closer I moved into the 'leopard crawl', edging forward on my tummy to peer over the crest. Below me on the beach were perhaps 1,000 Atlantic grey seals, hauled up on the warm sand. There were lots of pups, old enough to be gambolling in the shallows, nipping each other in play, barking, yelping, and annoying their parents. It was a magnificent sight – I can think of very few occasions when I have seen such numbers of truly wild large animals anywhere in Britain. Better still, they were entirely unaware of my presence, and I spent ages photographing them with my camera, though the range was a little too distant. I lost track of the time, and eventually had to sprint back across the wet sand, splashing through the shallow water as the tide came in, to get back to the island where our boat was pulled up.

It had been a fascinating day, but it held one more surprise for us during our return journey. Despite the wind and spray I was nodding off in the boat when it suddenly veered to one side, almost causing me to fall out. As I grabbed the rope along the rubber wall of the boat, a huge grey-brown fin passed within touching distance, protruding perhaps a metre from the sea, followed by a second, much smaller one. The dorsal and tail fins of a basking shark: Craig had swerved to miss it. Sadly the boat frightened it, and though we circled back to get a better look it had sunk beneath the choppy waves. Harmless filter feeders, basking sharks are the second biggest fish in the sea, growing up to ten metres or more. Sadly, these slow-moving giants have been hunted to extinction in many parts of the world, but here in the Hebrides there are still reasonable numbers.

The future of this remote corner of the UK, and of the wildlife it supports such as the great yellow bumblebee, is far from secure. Our studies of the genetics of small, isolated populations of bumblebees have found that these populations have little genetic diversity, exactly as we would expect. This means they have limited capacity to evolve if their environment changes, and they are likely to be susceptible to outbreaks of disease. It is very likely that this same issue of inbreeding affects many of the other organisms that live here, for most inevitably survive as small, isolated populations. Once upon a time, when these species were also found on the mainland in abundance, extinction of island populations could be reversed by re-colonisation, and genetic diversity could be boosted by occasional immigration. Now, for many Hebridean species, there is little chance of that. They are on their own.

Worryingly, this unique environment faces many other challenges. I have already mentioned the changes to crofting that are taking place, and the depopulation of these islands. One might think that fewer people would be good for wildlife, but much of the floral diversity of the machair is the result of human activity – the small-scale, low-intensity farming. The move towards large-scale sheep ranching is not favourable for bees or flowers – as we saw on the Monachs, too many sheep can be a disaster. At the opposite end of the spectrum, land abandonment can be almost as bad.

In the longer term, there is a potentially more devastating threat, in the form of climate change. The great yellow is a species well adapted to cold climates. It is big, with a long, shaggy coat, both attributes that help it to keep warm in the cool, damp climate of north-west Britain. It was always most common in the north, and as it declined in Britain in the face of loss of flower-rich habitats, its range contracted northwards to the places to which it is best adapted. It cannot go any further north. A warmer climate is likely to favour the displacement of the great yellow by more southerly species moving northwards. Since there is nowhere for the great yellow to be displaced to, that could be the final nail in the coffin for this species in the UK. Climate change also poses another threat. The machair is just inches above sea level, and thus even small rises in sea level – as predicted in the coming decades – might mean that the machair becomes inundated by the sea. This is particularly likely if extreme climatic events such as storms become more common – as they are predicted to do, and as we have already seen in recent years – because these can rip holes in the protective coastal dunes, allowing the sea to burst through. Of course, if this were to happen, then more or less all of the species that live here would be lost, not just the bumblebees, and there is very little that anyone could do locally to combat this.

Perhaps the best long-term hope for the great yellow in Britain is to work on boosting its populations on the Scottish mainland. There is a string of small populations clinging to the northern coast of Caithness and Sutherland, and here at least they are not going to be inundated by the sea as the coast is mostly rocky and steep. The Bumblebee Conservation Trust has done a lot of work in the area, encouraging farmers and working with local communities to sow flower mixes and rotate grazing to allow pastures to flower. The Royal Society for the Protection of Birds are also busy, trying to create and manage habitat for both corncrakes and great yellows at Durness, Dunnet Head, and elsewhere. Corncrake numbers are tiny but increasing. The extent of the range of great yellows on the north coast seems to have increased slightly, though it is unclear whether this is due to an increase in recording efforts or to a genuine spread of the bee.[10] Let's

[10] The Bumblebee Conservation Trust's former Scottish Officer, Bob Dawson, made a hobby out of attempting to locate great yellow bumblebees in new ten-kilometre-grid squares, targeting those adjacent to known populations and scouring any suitable habitat. If one browses the online distribution maps for the great yellow available on the National Biodiversity Network webpage, it appears to have almost doubled its range in Northern Scotland in 2008–10, but I fear that this is largely an artefact of Bob's efforts. There is a more general point that it can be very hard to disentangle genuine range changes from changing recorder efforts.

hope it is the latter. Realistically, there is little chance that either species will ever reclaim their former range, and most of us will rarely if ever get to see or hear them, living as they do in such remote places.

It is often argued that the best way to justify conservation efforts is to calculate the value of the services that wildlife provides – in the jargon of the day, we calculate the 'ecosystem service' that they provide. Bees are often used as an example; pollination is worth perhaps $215 billion globally, calculated as the value of crops we would lose without bees and other pollinators. It thus makes sense to look after them, because they look after us. The same argument can be rolled out for many other organisms; ladybirds and hoverflies that eat greenfly, flies that decompose dung, and so on. It is an argument that does not sit well with me. Why should nature only have worth if it does something for us? Just how self-obsessed are we? In any case, this approach falls flat on its face when presented with the corncrake or great yellow bumblebee. Recent studies suggest that most crop pollination is delivered by just 2 per cent of the bee species in an area – the common ones, as you might guess. Economically, great yellow bumblebees are utterly inconsequential; they probably contribute a tiny bit to the pollination of a few garden vegetables on crofts in the Uists. In truth neither corncrakes nor great yellow bumblebees – nor a great many other plants or animals – contribute anything meaningful to ecosystem services – we would not be noticeably worse off in any easily measurable way if they were one day to shuffle off – but I for one would consider that a very sad day indeed. We should not ask what nature does for us, but instead ask what we can do for it. I don't know when I will next get a chance to go to the Hebrides to see these fascinating creatures, but the world is a richer place for their being there.

CHAPTER THREE

Gorce Mountains and the Yellow Armpit Bee

Look deep into nature, and then you will understand everything better.

Albert Einstein

My travels in search of rare bumblebees have taken me to many corners of the UK over the years – from the Scottish Highlands in search of montane species to the iris-lined ditches of the Somerset Levels via the coastal dunes of Pembrokeshire, the Fens of East Anglia and the rocky cliffs of Cornwall. In almost all of these places – the sole exception being the Outer Hebrides – the nationally rare bees remain very hard to find, even in the most spectacularly beautiful of habitats with more flowers than one could shake a stick at. Yet the old books on bumblebees suggest that these species were once quite common, if never as abundant as the ubiquitous species such as the buff-tailed bumblebee. In Frederick Sladen's *The Humble-Bee* from 1912, he talks about shrill carders and short-haired bumblebees (the latter now extinct in Britain) as if they were very familiar creatures to him. The short-haired bumblebee he describes as 'common in many localities in the south and east' and the shrill carder as 'common in a good many places'. The ruderal bumblebee was in his day known as the 'large garden

49

bumblebee', suggesting that it was regularly found in gardens. He found and dug up the nests of all of these species and describes them in detail – whereas in over twenty years of studying bumblebees I have never found a single nest of any of these species, even with the help of a succession of sniffer dogs that we had trained by the army specifically to find bumblebee nests.[11]

It seemed to me that the changes that have occurred to the landscape in Britain are so profound that, even in the relatively unspoiled fragments of habitat, perhaps all that remains is a pale shadow of their former natural glory. It is hard to know for sure. Without a time machine, we can never really know what it would have been like to be a naturalist rambling through the British countryside in the eighteenth or nineteenth centuries, other than by reading their books and notes. We might infer from the fact that there are old recipes for cowslip wine – which require as the first step the collecting of two buckets full of cowslip flowers – that they were once much more common than they are now, but we can't know how common, or how abundant the bees were that visited them, or how numerous the worms that burrowed beneath their roots. This said, it occurred to me that there might be a way of gaining an insight into what Britain used to be like – by going to Eastern Europe. I had heard that in parts of Eastern Europe agricultural systems remained little changed, having escaped the drive for increased yield that afflicted Britain from the Second World War onwards, and which was subsequently driven throughout Western Europe by the Common Agricultural Policy's labyrinthine and often perverse system of subsidies for farm 'improvement'.

[11] Read *A Sting in the Tale* to find out more about the adventures of Toby the bumblebee sniffer dog.

Poland seemed like a good place to start. Perhaps there I might find species such as the shrill carder in abundance, sufficient that we could gather lots of data on their habits, favoured foods, preferred nesting places, and so on. Old museum records suggested that the rugged mountains on the borders of Poland, Slovakia and the Czech Republic were extraordinarily rich in bumblebees, with many species that had never occurred in Britain. The prospect of potentially seeing exotic species such as *Bombus veteranus* and *Bombus wurflenii* was mouth-watering.[12] Friends returning from holiday reported seeing farmers still using horses to pull their carts, and old-fashioned handmade haystacks in the fields. It sounded as if farming here might still be much as it was in Britain one hundred years ago, the nearest thing to a time machine that I was ever likely to find. And so it was that in August 2006 I found myself on a cheapo-jet flight to Krakow with Ben Darvill and Gillian Lye, the latter having just started a PhD with me on the nesting biology of bumblebees. We didn't tarry in Krakow, despite its famed beauty and charm, for we were here to see bees – we hired a car and headed due south from the airport, towards Zakopane in the foothills of the Tatra Mountains.

The Tatras form the western end and the highest part of the 1,200-kilometre-long Carpathian mountain range, which extends from Poland south-east through Slovakia, Ukraine and Romania. On the Polish side they are a popular skiing destination in winter and favoured by hikers in the summer, though attracting mainly

---

[12] Of course these bees have no common English names as they have never occurred in the UK, but one can have fun inventing suitable names for them. I'd suggest the old carder bumblebee and the red-tailed robber bumblebee for these two, the latter for reasons that will become clear.

Polish holidaymakers rather than those from further afield. The Tatras are quite recently formed by geological standards – along with the Alps, they were thrown up when Africa, rushing head-long in a northwards direction at the approximate speed of a growing fingernail, crashed into Europe, a slow-motion collision that began about 100 million years ago and is still happening today. Being young, the Tatra Mountains have not had long to erode, and so they are steep-sided with jagged peaks – very reminiscent of the Alps, but a little smaller, with the highest peak being at 2,650 metres. We found a small hotel in Zakopane and that evening we dined heartily on sausages and *pierogi* (dumplings),[13] on the pretext that we would need plenty of energy for hiking up the mountains in search of bees the following day.

The morning got off to a bad start. We drove out of Zakopane on the only road that appeared to head into the mountains, failing to notice a tiny sign that said 'no entry' in Polish. At the end of the road was a policeman, whose job it was to fine the succession of unobservant tourists who drove up. Our wallets duly lightened, we drove back to Zakopane and caught the tourist bus, which took us back past the policeman – who still seemed to be doing good business – to the bottom of the mountains. On the Polish side the Tatras are a national park, and we had to pay a hefty entrance fee to enter, but at least when we began to ascend the steep mountain trail we were not unnecessarily burdened with cash.

The path initially wound through dense forests, a mix of dark conifers and deciduous trees, with very few flowers or insects of any sort in evidence. After a couple of hours of steep ascent we

---

[13] I greatly enjoyed Polish food, but by the end of the trip I had had quite enough of their sausages, which came in an astonishing variety of shapes, colours and sizes, but all tasted much the same to me.

emerged into an alpine meadow, and suddenly there were flowers and bees everywhere. Dramatic indigo spikes of monkshood rose amongst the yellows of St John's wort and saxifrage, mauve powder-puffs of scabious, and the delicate blue bells of campanulas. On the more disturbed, steeper slopes where rock falls had recently occurred, swathes of rosebay willowherb clothed the mountainside in pink. We scrambled about – trying to catch bees on such steep meadows was challenging as one couldn't keep an eye on the bees while watching one's footing, so all three of us fell over a few times in our excitement. Ben and I were old hands at catching and identifying bumblebees in the UK, whereas Gillian was pretty new to the whole thing, but pretty soon all three of us were confused. Compared to the UK, we seemed to be encountering a bewildering array of species: some looked familiar, others were similar but slightly odd, and others still were quite unlike any UK species. We'd known to expect this and had tried to prepare, but nonetheless it was clear that it was going to take some time to get to grips with all of these new bees. We took lots of photographs, and some voucher specimens so that we could confirm our identifications.[14]

Over the next few days we hiked up and down the mountains, collecting a raft of records of the numbers of different species, and of the flowers they were feeding upon. The species we had never seen before included *Bombus wurflenii*, a handsome beast resembling the much more familiar red-tailed bumblebee, but with longer, shaggier fur to help it keep warm in these high mountains, slightly brighter markings, and a penchant for larceny which led

---

[14] It is, regrettably, sometimes necessary to collect insects if one wishes to know for certain what species they belong to. Unless one can identify them with certainty it is not possible to study them, or work out how to conserve them.

us to christen it the 'red-tailed robber bee'.[15] The monkshood flowers have evolved to be pollinated by long-tongued bees, with their nectar hidden at the end of a long, curved tube, but the red-tailed robber bees unceremoniously bit a hole in the side of the flower using their especially sharp, toothed mandibles, and stole the nectar. There was also the 'Pyrenean bumblebee' (*Bombus pyrenaeus*), an endearingly pretty little bumblebee, resembling its close cousin which we know in the UK as the early bumblebee, but again with a very fluffy, long coat and more yellow. This is a high-mountain species, as you might guess from the name, and it absolutely loved the willowherbs. We also saw a few *Bombus veteranus*, the 'old carder bumblebee', a brownish bee quite similar to our common carder, but with a dusky, smoked colouration.

Conspicuously absent from the fauna were the rare British species that I had hoped to find out more about – there were no shrill carders or ruderal bumblebees, for example, though I knew that they ought to be in southern Poland. This wasn't really surprising – we were in the high mountains, and so were mainly finding alpine specialists, not species that one would expect to see in Britain. These high-mountain meadows were beautiful, but they didn't provide the window on the past that I had hoped for. So, after a few days of trekking up and down the steep slopes of the Tatras, we decided to head elsewhere.

We had no particular basis on which to make a decision as to where to go next, so we arbitrarily set off from Zakopane in a north-easterly direction. It was initially a bit of a drizzly morning,

---

[15] This is a species I was later to come to know well during fieldwork in the Swiss Alps, as described in *A Buzz in the Meadow*. There we found that *B. wurflenii* have a habit of always robbing flowers on the same side, with individual bees copying each other's robbing behaviour right down to the detail of which side of the flower they attack.

and our route took us through rolling countryside, the land divided into long, thin fields, many of them only twenty metres or so wide and perhaps one or two hundred metres long, separated by grassy strips. This was nothing like farming in the UK – a tractor wouldn't even be able to turn around in many of these fields. Or, at least, it was nothing like *modern* farming in the UK – this was more like the medieval system of open-field farming, whereby the fields belonging to the local lord were divided into numerous strips, with each serf allocated a handful of strips to cultivate. Over centuries the movement of soil when ploughing these strips created furrows between them, and the ridges and furrows can still occasionally be seen in some British fields that have never been ploughed with modern machinery. We stopped and mooched around a bit, but the damp weather was keeping the bees at bay. However, the habitat looked promising – many of the strips were fallow, while others had been sown with clover leys,[16] red clover, a mixture of red, white and zigzag clovers, or sainfoin. Fallows and clover leys are both now very rare in modern farming, but were a part of crop rotations for hundreds of years before cheap artificial fertilisers became available and enabled farmers to grow arable crops every year, without ever resting the land.

A little later in the morning we passed through the depressingly drab town of Nowy Targ, which appeared to consist entirely of blocky, weather-stained concrete buildings from the communist era. Shortly afterwards we found ourselves on the southern edge of a range of low, rounded mountains, much less dramatic than the Tatras. These were the Gorce Mountains, a range barely mentioned in our guidebook, which briefly but intriguingly described them as a wild place where wolves, lynx and bears were

---

[16] 'Ley' is a term for resting the land, either by leaving it fallow or by sowing a legume crop and then ploughing it in to boost soil health.

to be found. The book didn't mention bumblebees – sadly, the bee section in most travel guides is woefully inadequate – but we figured that somewhere that was good for bears must surely be good for bees. By this time the sun was burning through a little so we pulled over, parking by a bridge over a bubbling stream that flowed south from the mountains.

We were not in the mountains proper, but in an area of rustic wooden farmhouses, with strip fields and small orchards, the latter mostly planted with old, gnarled apple trees crusted in lichens, proper full-sized trees rather than those with dwarfing rootstocks grown in modern orchards. It really was as if we had gone back in time one hundred years or more. Two aged men were cutting a wheat crop with scythes, then deftly stooking the cut crop by hand (stacking the stems into pyramidal mounds with the grain at the top) to allow the grain to dry and mature before threshing. I had never seen this before – it is a practice that probably went on for the best part of 10,000 years until modern machinery was invented with the intention of making such back-breaking work a thing of the past. Nearby, a woman was cutting fodder from the road verge with a sickle and using her apron as a bag to carry it, presumably back to some livestock. As I watched her, a cheerful-looking young guy called out a greeting to us as he trotted past on a horse-drawn cart laden with marrows, the pneumatic tyres on the cart the only slight concession to modernity. He stared curiously at our butterfly nets and other paraphernalia, so I mimed catching a bee, which probably only served to increase his bafflement. It clearly amused him anyhow.

We headed off in different directions in search of bees. The higgledy-piggledy nature of the landscape, where everything was done on a small scale, meant that there were lots of edges and forgotten corners where flowers could grow: 'waste places', as my flower guide describes them, though in my view they are far from being a waste. The crops themselves often contained flowering arable weeds – familiar species such as cornflowers, fumitory and poppies, and others I had never seen such as large-flowered hemp nettle, a splendid herb with yellow and purple tubular flowers that long-tongued bees clearly adored. It was evident that herbicides were not much used here. These same arable weeds also thrived in the fallow fields, just as they do on the crofts of the Outer Hebrides. As we had seen earlier, some of the field strips were sown with red clover leys, and they were a-humming with bees. Between the field strips, beneath the orchard trees and along the road and track verges, communities of perennial wildflowers had developed: knapweeds, woundworts, cat's ear, clovers, scabious, marjoram and thyme, along with wood cow-wheat, a most unusual plant in which the young leaves are vivid purple, which together with the yellow and red flowers creates a most exotic appearance. The banks of the stream were rife with the pendant pink blooms of Himalayan balsam, an invasive weed but much loved by bees.[17] Whichever way one turned, there were patches of flowers and the buzz of insect life.

In the clover leys there were garden bumblebees, alongside ruderal, great yellow and shrill carder bumblebees. In an orchard I found red-shanked carders, buff-tails, red-tails and tree bumblebees. Along the track verges were common carders, along with old

---

[17] Invasive weeds – non-native plants that run amok in the wild – are a major threat to biodiversity, and Himalayan balsam is one of the worst in Europe, smothering native stream-bank vegetation from Britain to Poland and beyond. If there were a way to effectively control it then I would have to agree that this would be a good thing, but there would undoubtedly be negative consequences for our poor bumblebees. In the flowerless void that is most modern British farmland, stands of balsam along ditches and streams are often amongst the only sources of food left for them.

carders and early bumblebees. There were other species too, once again some that I did not immediately recognise – bees that were entirely black with white tails, or black but with yellow tufts near the base of their legs, and still others with distinctive yellow stripes and white or orange tails. Most of the species that I couldn't identify were males, and I caught a selection for closer scrutiny. When we rendezvoused back at the car an hour later, between us we had recorded fifteen species of bumblebee that we could put a name to, plus what appeared to be several others. So that we could be consistent when we saw them again, we agreed on names for each of the bees that we couldn't identify – the black bees with white tails became, rather unimaginatively, 'black white-tails', and the ones with yellow tufts around the base of their legs became 'yellow armpit bees'. This may not sound very scientific, but it was the best we could do at the time.

We headed on into the Gorce Mountains, following a single-track road that wound along a long river valley. We stopped and searched for bees every few kilometres, spending an hour at each place, counting the numbers of each species and recording the flowers they were on, just as I had on Salisbury Plain. I have rarely explored anywhere so idyllic.

That evening we compared notes, and over some delicious and exceedingly cheap Tyskie beer, Gillian and I attempted to work out what all the different species were. As is often the case in entomology, in the end it all comes down to the genitals. I spent about six months during my PhD staring down a microscope at the genitalia of butterflies. Many species of insect can only reliably be distinguished by staring very hard at the male genitalia, which are usually of a unique and often bizarrely complex shape in each species. The female genitals tend to be of little help, but fortunately it was August and males were plentiful. We had no microscope, only a hand lens, and it was tricky, fiddly work trying to get a clear view of such tiny organs by the light of a dim lamp. The beer probably didn't help in this respect either.

Eventually we managed to sketch diagrams of the various species we had, and could then compare them to photographs I had brought with us of the genitals of all of the European species (so far as I know there is no law against smuggling bee porn across international borders, at least within the EU). We were in for a bit of a surprise. Gillian, a very pretty, quiet girl, who turned out to be the brains of the party, was the first to spot that all of the odd-looking males that we did not recognise were more or less identical in the tackle department, and their genitalia appeared to match perfectly the photograph I had of the genitalia of the broken-banded bumblebee. I had only ever seen this species on Salisbury Plain, and there all the males were very similar – and closely resembling buff-tailed bumblebees, with two yellow bands and a whitish tail. In Poland, it seemed that the males of this species had gone berserk, adopting a bewildering and rather attractive array of colour patterns. This is unexpected in bumblebees, for their coloured stripes are thought to have evolved as warning colours to signal that they have a sting, and hence keep predatory birds at bay (though great tits and bee-eaters wolf them down regardless). What is more, it has been suggested that the reason many species look so similar is that they mimic one another – by all adopting the same colour patterns, they collectively send a stronger signal to their predators.[18] Male bumblebees don't have

[18] Of course, no one is suggesting that the bees thought this through, got together at a big meeting, and collectively agreed on a common colour scheme. The idea is just that natural selection favoured individual bees that most closely resembled the colour pattern of whatever happened to be the most common species in any particular area. The phenomenon of stinging or poisonous species coming to resemble one

a sting (which evolved from the egg-laying tube, which of course males don't have) and so I presume that their coloured stripes are a bluff. By resembling the females, they may benefit by association – birds having learned earlier in the year from experience with the queens and workers that creatures with this colour pattern are dangerous.[19] One might therefore expect the males to look very like the workers, but in many species they do not. Usually they are a little more brightly coloured, often with broader yellow bands and fluffy yellow faces. This would seem to be a foolish strategy so far as the risk posed by predators is concerned, but is presumably the result of sexual selection – perhaps the virgin queens prefer brightly coloured males, and so they have to have bright colours to stand any chance of mating. Having sex and then being eaten is a success, in evolutionary terms, compared to living a long life without either. I must confess that this is all wild speculation – I have no evidence whatsoever that queen bees are attracted to brightly coloured males, but it happens in butterflies and birds, so why not bees? Actually, I did once have a Bangladeshi postdoc who studied mating choices in bumblebees and found that females were more likely to mate with the males with the longest legs, but that is another story.

Anyway, to cut to the chase, it is very hard to explain why male broken-banded bumblebees in Poland are so variable in colour. One might expect them to be brightly coloured to impress a mate, or to copy the colours of the females to avoid getting eaten, but

---

another is known as Müllerian mimicry after the German naturalist Fritz Müller who first proposed the concept.

[19] When a palatable species evolves to look like a poisonous or dangerous one it is known as Batesian mimicry, after Henry Walter Bates, who noticed that edible species of Brazilian butterfly often closely mimic poisonous species.

adopting wildly varying patterns makes no sense. Perhaps Polish females have eclectic tastes, or prefer to mate with unusual-looking males, which might favour the proliferation of colour forms. It would make a fun research project to investigate further.

At least now we were able to identify the different species that we found, and could set about collecting lots of data on the various types of bumblebee in the area. We settled in a hotel in Ochotnica Gorna, a picturesque village which straggles along the south bank of the Ochotnica River in the heart of the mountains. How the hotel could possibly make a living was unclear as it was almost empty in the height of the summer season. There seemed to be very few tourists here compared to the Tatras, and no overzealous traffic policemen, which suited us just fine.

We spent the next week exploring the rolling mountains and valleys in search of bees. Along the river valleys were strings of small farms, all with tiny fields, and minimal signs of farm machinery. Some farmers had small tractors, but there was no space for the huge combine harvesters that we see in the UK and so almost all of the harvesting was still done by hand, while horse-pulled carts seemed to be the most common form of transport. Higher up, the farmland gave way to pastures, grazed by small herds of cattle and sheep, and higher still were areas of dense forests interspersed with heathland. It was bilberry season, and we encountered many locals out collecting buckets full of the small purple fruits, some armed with odd contraptions consisting of an array of parallel metal blades fitted to a glove, with which they combed the berries from the bushes. When I first saw a guy emerging from the bushes, dark juice dripping from the spray of blades attached to his hand, my heart skipped a beat as I recalled the Freddy Krueger character from the horror film *A Nightmare on Elm Street*.

We didn't encounter any bears or wolves on our travels, sadly, but we did see a host of other wildlife. Butterflies were everywhere – blues, fritillaries, swallowtails, whites and browns – my favourite being the scarce copper which abounded in the mountain meadows, their red-gold metallic wings gleaming in the sunshine. The bumblebee-mimicking hoverfly *Vollucella bombylans* was common, wonderfully furry flies that come in a range of different colour morphs, each copying the patterns of a particular bumble-bee species. I saw my first wart-biter cricket – a huge and beautifully camouflaged emerald-green and black beast which lurked amongst the tussocky grass on south-facing slopes. In the UK this species occurs at just three sites in the South Downs near Brighton, but in the Gorce Mountains it seemed quite common. They have fearsome mandibles, from which they derive their name – in days gone by, it is claimed that they were used in Sweden to bite off warts, though it would seem to me that there must be easier ways of doing this. We didn't find many more bumblebee species to add to the tally from our first day – a handful of cuckoo bumblebees, including one new one for me, *Bombus quadricolor* (the 'four-coloured cuckoo bumblebee'?), which specialises in attacking the nests of the broken-banded bumblebee. What we did find were a great many of some of the species that are enormously rare in Britain – just as I had hoped. Shrill carders were common everywhere, alongside red-shanked carders, broken-banded and ruderal bumblebees, plus all of the common British species.

So what did all our data show? Since I began searching for rare bumblebees on Salisbury Plain in 2002, I had been trying to work out why some species have declined precipitously, in some cases going extinct, while others seem to be doing pretty well. This is actually a general question one could apply to most groups of animals or plants. Why are marsh tits less common than blue tits? Why are lady's slipper orchids close to extinction, while the common spotted orchid remains, as the name suggests, common?

If one studies organisms in sufficient detail, one may discover the answer. For example, researchers have found that the Adonis blue butterfly is warmth-loving and can only thrive on close-grazed, south-facing chalk downland and so it is inevitably scarcer than its cousin the chalkhill blue, which is slightly less fussy and can live in shadier, longer turf and north-facing sites. Bumblebees had thus far proved to be more reluctant to give up their ecological secrets.

Once back from Poland, I spent many hours analysing the huge data set that we had gathered. Crunching data is not everyone's cup of tea, but it can be surprisingly satisfying, especially when clear patterns emerge. Most of the species that have become rare in the UK had rather similar flower preferences – they tend to be longer-tongued bees, very partial to red clover and other legumes with deep flowers, and also to plants in the mint family such as the hemp nettle. The broken-banded bumblebee was the only exception – this is a short-tongued bee, and in Poland it visited a huge range of different flowers. While I was playing around with these data, another of my PhD students, Claire Carvell, showed me her analyses of changes in abundance of flowers visited by bumblebees in the UK between 1930 and 1999. Her studies neatly but depressingly showed that 76 per cent of flower species had declined in the frequency with which they were recorded in survey plots, but that deep flowers seemed to have been particularly hard hit. Red clover had declined by 40 per cent in frequency in just twenty years, from 1978 to 1998. It was hardly surprising, then, that the long-tongued bees, which in Poland were still thriving in the clover leys, have declined in Britain. There is still red clover about, but there is far less of it than there used to be.

At more or less the same time that Claire and I were comparing notes, two Dutch biologists, David Kleijn and Ivo Raemakers, were trying to tackle the same question, and had hit upon an ingenious idea as to how they might travel back in time to study what bumblebees used to do. Rather than going to Poland, they went to the museum. The Netherlands has a similar history of agricultural intensification to Britain and is also a very crowded country, so its bees show a similar pattern of decline. In fact, exactly the same three species that went extinct in the UK also disappeared from the Netherlands (the apple bumblebee, Cullem's bumblebee and the short-haired bumblebee), and a near-identical set of species have declined dramatically, including the brown-banded and shrill carders, and the ruderal and broken-banded bumblebees. Europe's museums are stuffed full of pinned bumblebees, many caught during the great era of amateur insect collecting in the first thirty years of the twentieth century. Klein and Raemakers realised that many of these bees still had pollen in their pollen baskets, pollen that had been sitting there for roughly one hundred years. It is one of those brilliant ideas that seems obvious once you've thought of it – by identifying the pollen grains, they could quantify what flowers these bees were visiting all those years ago. They collected and analysed the pollen from the legs of museum bees, not just from the Netherlands but also from Belgium and the UK. They then went back to the same sites where those bees had been caught, and collected fresh pollen from the common species that still lived there (the declining species had gone from the sites, though they cling on elsewhere). Their data confirmed that the species that were subsequently going to decline tended to collect pollen from fewer plant species than those that managed to adapt to the coming changes. They were more specialised – several of them, particularly the long-tongued species, were very dependent on red clover and other legumes. The broken-banded bumblebee seemed to be heavily dependent on harebells, while the heath bumblebee favoured heathers and bilberries. In comparison, the species that coped better with what the twentieth century was to throw at them tended to have more catholic tastes to start with, visiting a broad range of flowers for many different plant families. What is more, they found that the flowers that the declining species preferred had themselves declined more than average during the twentieth century – these bees had been unlucky, happening to prefer plants that weren't going to do well. Harebells, for example, are now pretty scarce flowers in most places, so any bee foolish enough to have a real taste for them was going to be in for a tough time. This of course was more or less exactly what our own data showed – that our rare species had become rare because their preferred food had become scarcer. In Poland, still rich in flowers of all types, there was room for all of them to thrive. It is always reassuring when studies addressing the same question from completely different angles arrive at more or less the same conclusions.

Our bumper data set of bee visits to flowers also allowed me to investigate a different question, relating to competition between bumblebees. When I was an undergraduate, I was taught about a classic study conducted on the bumblebees of Colorado in the 1970s by an Australian named Graham Pyke. He hiked up and down mountains counting bumblebees in a rather similar way to our surveys of the Polish mountains, but he had a different interest. He was looking for evidence that bumblebees compete with one another. Bumblebees of different species all tend to be pretty similar in shape, size and in their basic biology. Of course some are a little smaller or larger, or more or less furry, and they differ in colour, but they are all flying about together feeding on broadly the same resources – flowers. Back in the 1930s, a Russian ecologist named Georgy Gause had proposed the 'competitive exclusion principle', which eventually became known as Gause's Law. This states that two species that use exactly the same resources (which might be food, nest sites or anything else needed for survival) cannot coexist. The point is that whichever is the superior competitor will win the competitive battle, and exclude its rival. He demonstrated this in lab experiments using two species of *Paramecium*, microscopic freshwater protozoans related to *Amoeba*. In any particular conditions of water quality and food availability one or other of his types of *Paramecium* would always eventually win out. Bumblebees seemed like an interesting system to see whether such things happened in the real world, and this is what Pyke set out to discover.

Pyke gathered data on which species of bumblebee occurred at different altitudes as he climbed up each of a number of mountains, and he discovered that there were only ever three or four species found in abundance in any particular mountain meadow. When he looked at them more closely, he found that there was always one short-tongued species, one with a medium tongue and one with a long tongue. As you might expect, the short-tongued bees fed on shallow flowers, the long-tongued bees fed mainly on deep flowers, and the bees with medium-length tongues fed mostly on flowers that were in between, so that the three species carved up the available flower resources between them and avoided competition. The particular species that were present changed with altitude, and sometimes from mountain to mountain. In addition, a fourth species, the western bumblebee, was also present at most sites, but this is a nectar robber like the red-tailed robber bee in Poland. Moreover, the western bumblebee specialised in stealing nectar from very deep flowers normally visited by hummingbirds, so it didn't compete at all with the other species. Pyke's data suggested that Gause was correct – bumblebee communities were structured by competition, so that species with similar tongue lengths did not coexist. Only species that differed in tongue length – and so tended to visit different flowers – could thrive together. One can make parallels with Darwin's finches, where the different species have evolved different shaped bills to carve up the resources between them and minimise competition – some have long thin bills for gleaning insects, others slightly fatter bills for crushing small seeds, and still others have fat bills for crunching big seeds and nuts.

Our Polish data didn't seem to fit Gause's Law. We had found up to fifteen species living alongside one another, sometimes in a single meadow. Admittedly, for the long-tongued bees only one species, the garden bumblebee, was usually common, with just a few ruderal bumblebees turning up alongside them. Similarly, the common carder was usually by far the most abundant of the medium-tongued bees, even though there were quite a few shrill, red-shanked and brown-banded carders about. But amongst the short-tongued bees there were often four or five common species living together – buff-tails, white-tails, early, tree and broken-banded bumblebees. According to Gause, one species should have outcompeted all the others. As they say in the film *Highlander*, 'There can be only one.' So what was going on?

When I looked more closely at the flowers that they were visiting, it turned out that these short-tongued bees actually did tend to visit different flowers, though there was quite a bit of overlap. White-tailed bumblebees were often on umbellifers (hogweed, angelica, and so on), which the others seemed to avoid. The tree bumblebees went mad for willowherb, while the broken-banded bumblebees were very keen on knapweeds (we saw very few harebells in Poland, the flower they had historically favoured in the Netherlands, UK and Belgium). Even though the different species are equipped with near-identical mouthparts, the bees were somehow dividing up the resources and minimising competition. Quite how they arrived at this equitable distribution of resources we don't know. We also have no idea why this doesn't happen in Colorado. It could possibly be because bumblebees have lived in Europe for thirty to forty million years, while they have only been in America for about twenty million years, so perhaps they have had more time to evolve into more specialised niches here. I don't find this idea particular convincing, since twenty million years is surely quite a long time in anyone's book, but it is the best idea that I have managed to come up with so far. There is so much that we still do not understand about the way communities of animals and plants interact with one another.

These comparisons highlighted just how rich the Polish bumblebee community is compared to what remains in the UK. Of course the Gorce Mountains are far from an exact replica of what Britain used to be like before intensive farming, but they perhaps give some idea. I think that Frederick Sladen would have felt quite at home, at least as far as the abundance of different bees and flowers are concerned. I'm not so sure how he would have felt about all the sausages.

I would not wish to condemn the farmers of this region to a life of back-breaking labour, but I hope that this backwater of rural life has remained unchanged since I visited. It may sound patronising, but there was no obvious poverty, and the people we met seemed happy enough. Like crofters in the Uists, they live in a beautiful, unspoiled place, eating healthy food, much of it grown by themselves or their near-neighbours. Are they better or worse off than those of us who spend our days commuting through traffic to work, then sitting at a workstation or in a meeting while daydreaming of our annual holiday to a Greek island? I don't know the answer to that question, but purely from the perspective of the bees I hope things have not changed, though I fear that change is inevitable. Just as in the Outer Hebrides, I wonder if young folk will be content to take over their small family farm in rural Poland. What's more, major external factors are at play – Poland joined the European Union in 2004, two years before our visit.

The Common Agricultural Policy must be a contender for one of the most unfair, opaque and perverse bundles of legislation ever concocted by man, and I fear that it may bring an end to the extraordinary biodiversity that has until recently thrived in parts of Poland and elsewhere in Eastern Europe. The CAP was introduced in the late 1950s, with the primary goal of increasing food production through subsidies for modernisation of farming and guaranteed prices for crops, regardless of demand. This goal seems laudable, particularly in the context of a Europe that had seen nearly two decades of food rationing and shortages brought on by the Second World War. The downsides of the policies were that we subsidised the destruction of the European countryside to make way for industrial farming, created huge surpluses of food, flooded the world market with cheap, subsidised produce, and gave EU farmers an artificial advantage over farmers in developing countries, thus condemning millions of farmers elsewhere in the world to a life of poverty. The CAP has been revised countless times, but has not fundamentally changed. In essence, it still involves taking staggering amounts of EU taxpayers' money and using it to subsidise industrial farming, giving the bulk of it to multinational farming corporations and Europe's major landowners, all of whom are already pretty well off by anyone's standards, while giving tiny amounts to small-scale farmers, those striving for more sustainable food production and farms in marginal areas – the ones that might justifiably deserve our support. Land-grabbing – the buying up of large tracts of farmland by multinationals, usually accompanied by the displacement of local people and industrial-scale farming – is something associated mainly with Africa and South America, but in recent years it has begun across Eastern Europe, particularly in countries such as Romania, Serbia, Hungary and Ukraine. In Poland, purchase of land by foreign corporations requires a special permit, but this law expired in 2016 so it will be possible for multinational farming corporations to buy up land, which remains cheap, and roll out their industrial model of farming, while being heavily subsidised by us taxpayers. In February 2015, Polish farmers used their tractors to block roads across the country in protest at these threats to their livelihoods, but whether they can hold back the tide of 'progress' remains unclear.

There is a positive side to this story. The fact that taxpayers' money is used to subsidise farming across Europe means that taxpayers have every right to demand a say in how that money is spent. With a major overhaul to redirect funds from large corporations to small-scale farmers, to encourage sustainable, environmentally friendly farming practices and local food production, the Common Agricultural Policy could make life better for farming communities, rather than subsidising their destruction. Of course it is not easy to make such major changes, but that is no reason not to try. Perhaps that is a story for another day, or even another book.

CHAPTER FOUR

Patagonia and the Giant
Golden Bumblebee

Consider the bee
Is it aware of its mortality?
Does it fear surprise?
Does it see many ghosts with its many eyes?

Mal Campbell, from 'Worship the Ant'

On 1 January 2012 I flew from Glasgow via Heathrow and Frankfurt to Buenos Aires, accompanied by Jessica Scriven, a PhD student of mine who was studying the population genetics and ecology of the cryptic bumblebee[20] in the UK. The weather in Scotland, where we were based at the time, had been decidedly *dreich*[21] through December, so we were both looking forward to a few weeks of

[20] An aptly named bumblebee, it was only discovered in the UK in 2002 for the simple reason that it is indistinguishable from the white-tailed bumblebee unless one takes the trouble to look at its DNA or analyse its sex pheromones. As a result, we know almost nothing about this species, a situation that Jess is attempting to remedy.

[21] If you've not come across it, *dreich* is a Scottish colloquialism for damp, drizzly, cold and miserable weather. It is a useful term to describe the Scottish weather through much of the autumn, winter and spring, and often comes in handy in the summer too.

southern-hemisphere summer sunshine. Our self-imposed mission was to investigate the impact of European buff-tailed bumblebees on the native fauna and flora of South America. Rumours suggested that something pretty disastrous was afoot, something that threatened the future of all of the bumblebees of South America, and I wanted to see first-hand what was going on, and whether there might be any possible solution.

South America is the only part of the southern hemisphere to have naturally occurring bumblebees. Most bumblebee species live in the northern hemisphere; we think they originated somewhere in the eastern Himalayas some forty million years ago, and they spread through the temperate old world, west to Poland and eventually to Benbecula, east to Siberia and then across the Bering Strait to Alaska and hence to the rest of North America. With their large size and big furry coats, they tend to overheat in warm climates, so they never ventured south towards the Equator except in the Americas, where there is a more or less continuous chain of mountains linking the Rockies down through Central America to the Andes in South America. A few especially adventurous bees made their way down through the cooler habitats provided by these mountains, reaching South America somewhere between four and fifteen million years ago. They must have liked what they found, for they diversified, and today South America supports twenty-four known native bumblebee species. They include some really odd bumblebees, such as *Bombus atratus*, which has managed to adapt to live in the steamy tropical lowlands of South America, not a habitat one normally associates with bumblebees. It is an entirely black bumblebee with dark, smoky wings that superficially resembles a carpenter bee, and it is reputed to be absolutely ferocious in defence of its nest. Its life cycle is quite different to 'normal' bumblebees, for colonies can last for several years, there being no need for a winter diapause. There can also be up to eight active queens in a single nest, though apparently they fight a lot and eventually all but one are killed. Just as in Europe, some South American bumblebees have undergone major declines, most likely driven by intensive farming. For example *Bombus bellicosus* (the bellicose bumblebee?), a bee of open grasslands in Brazil, Uruguay and northern Argentina, is believed to be extinct in much of its former range. It used to be found in the area around Buenos Aires, but seems to have disappeared, and there have been very few records from anywhere in the last twenty-five years.

There was one particular bumblebee that I really wanted to see: *Bombus dahlbomii*, claimed to be the largest bumblebee species in the world, a creature that would make the great yellow seem like a mosquito by comparison. What is more, this wondrous beast was said to have a magnificent golden pelt (I can't quite bring myself to use the term fleece for the fur of a bumblebee). Such is their size, the queens are said to resemble flying golden mice, their generous proportions helping them to keep warm in the bleak, windy climate of Patagonia and Tierra del Fuego. This is the only native bumblebee found in the southern half of Argentina and Chile, and it is a creature strongly associated with the great mountain chain of the Andes. Unfortunately, local scientists were reporting the rapid demise of this splendid species following the invasion of South America by European bumblebees.

Man's enthusiasm for taking bees from one part of the world and releasing them in another appears to remain unchecked, despite the well-documented catastrophes that have occurred due to other invasive species running amok, such as mink and grey squirrels in the UK, or cane toads and rabbits in Australia. We have spread honeybees, natives of Europe and Africa, to every country in the world excluding Antarctica, and we have also deliberately carried numerous other bees, including various bumblebees and a host of small, solitary bee species, from one continent to another. In the late 1980s, Chilean agronomists imported ruderal bumblebees from

New Zealand to help with clover pollination. Ruderal bumblebees are not natives of New Zealand, but were taken there from Kent in 1885 for precisely the same reason; New Zealand farmers had noticed that their red clover set no seed, and had worked out that this was because there were no bumblebees to pollinate it. Just as in Poland today, the New Zealanders of the 1880s and Chilean farmers of the 1980s used clover as a ley crop to improve soil health. It is not clear why the Chileans were unhappy with the pollination provided by the native *dahlbomii*, but perhaps they were too few in the lowland areas where clover tends to be grown – *dahlbomii* is more a species of the mountains and cool, wet forests. Whatever their reasoning, ruderal bumblebees were shipped in with little thought for the consequences and soon became established.

Chile is an oddly shaped country, an immensely long sliver of land more than 3,000 kilometres from north to south, squeezed between the Pacific to the west and the Andes to the east. It stretches from the Atacama Desert in the north, one of the driest places on Earth, to rain-sodden Tierra del Fuego in the south, the closest land mass to Antarctica. The ruderal bumblebees were introduced near the capital Santiago, which is very roughly in the middle of Chile, and from there they spread both north and south, eventually being halted by the two climatic extremes.

In 1994 the first ruderal bumblebees were recorded in neighbouring Argentina. If your geography is a little rusty, Argentina lies east of the north-south chain of the Andes. It is as long as Chile, but roughly triangular in shape, widening from Tierra del Fuego (ownership of which is uneasily split between Argentina and Chile) to a rather bulbous mid and upper section which borders Brazil, Bolivia, Paraguay and Uruguay. Like Chile, the climate varies enormously from the icy, windswept south to high-mountain deserts in the north-west and subtropical forests in the north-east. The Andes are the second highest mountain range in the world, providing a formidable barrier between Chile and Argentina for most of their length, which is fortunate for there is little love lost between the two countries. However, near San Martin de los Andes in Argentina they are lower, with passes through the mountains no higher than about 700 metres. It is through these passes that ruderal bumblebees are thought to have flown, and soon they became abundant in the lush, temperate forests of southern beech and araucaria (monkey puzzle) surrounding San Martin. For a number of years they seemed to live alongside *dahlbomii*. Both are long-tongued species, so they visit similar, deep flowers, and there was concern that the two species might have read about Gause's Law in an ecology textbook and worked out that they should not be able to coexist. If ruderal bumblebees proved to be the superior competitor, perhaps they would wipe out the native *dahlbomii*. In particular, both bee species loved to feed upon the beautiful scarlet and purple flowers of the native fuchsias. Carolina Morales, a scientist based at the Universidad Nacional del Comahue in San Carlos de Bariloche, just to the south of San Martin, has been studying bumblebees in the region since the 1990s. She has counted numbers of different bumblebee species over many years and came to the conclusion that *dahlbomii* was a little less common following the arrival of ruderal bumblebees, but that there had been no dramatic impact. For a while, all was well – the two bumblebees seemed blissfully unaware of Gause's Law, and were able to get along with one another without disaster. Unfortunately, that was not the end of the story.

Buff-tailed bumblebees, familiar and beloved natives of Europe which were recently voted the UK's favourite insect in an online poll, have also been moved to far-flung lands, even more so than ruderal bumblebees. Buff-tails are widely used for tomato pollination, and are bred in huge numbers – about two million nests per year – in factories in Europe from where they are distributed all over the world. Along with ruderals, they were introduced from Kent to New Zealand in 1885 to help with clover pollination, and in 1991 they spread to Tasmania, probably with a little unofficial help from a tomato grower. In about 2005 they accidentally escaped from tomato glasshouses in Japan and are now thriving in the wild and spreading, causing great concern to conservationists. Potentially most disastrously, the Chileans decided to get in on the act and so they also imported buff-tails in 1998, presumably not satisfied with having just one non-native bumblebee species. It isn't clear what role the buff-tails were supposed to perform in Chile – they are no good at pollinating red clover since their tongues are too short, but perhaps they were intended to pollinate tomatoes.

Whatever their intended purpose, the buff-tails didn't stick around to find out. They are highly adaptable creatures, without doubt the most roughty-toughty European bumblebee species, able to survive in almost any habitat, including the most intensively farmed land, and naturally ranging from the semi-deserts of Morocco in the south to Norway in the north and Israel to the east. They rapidly came to terms with their new environment in Chile and set out to explore, multiplying as they spread. Buff-tails are a short-tongued bumblebee, so they tend to feed on rather different, shallower flowers than the long-tongued ruderal bumblebees and *dahlbomii*. These flowers are naturally pollinated by many hundreds of native Andean solitary bee species, mostly smallish creatures with short tongues about which precious little is known, and which may or may not be suffering from the presence of this new competitor. Buff-tails are also reasonably adept at nectar-robbing, stealing nectar from deep flowers when the need arises by biting holes in the back or side of the flower, so there are few flowers from which they cannot feed.

In 2006, eight years after they were introduced to Chile, the first buff-tails arrived in Argentina, following the trails of the ruderals through the mountain passes to San Martin. They quickly became enormously abundant, and to Carolina's consternation the other bumblebee species both more or less immediately disappeared. *Bombus dahlbomii*, once the only bumblebee in the area and a common sight in summer, seemed to vanish, along with the vast majority of the ruderal bumblebees. Around San Martin and Bariloche, this has remained the case ever since. Carolina has seen just a handful of *dahlbomii* in the last nine years, and fears that they could go locally extinct at any time.

It was tales of this dramatic decline in the abundance of the world's largest bumblebee that drew me to Argentina. There were major questions that seemed to be unanswered. What was causing the decline of *dahlbomii*? Was it competition with the invader, or was something else going on? How far had buff-tails spread, and were they wiping out native bumblebees wherever they arrived? Could anything be done to salvage the situation?

So it was that, on the morning of 2 January 2012, Jess and I arrived in Buenos Aires, jet lagged and bleary-eyed. The ancestors of *dahlbomii* had arrived in Argentina perhaps ten million years ago, the ruderal bumblebees eighteen years before us, while the buff-tails had been here for just six years at the time. We hired a car and set off to see which bees we could find. Buenos Aires is on the east coast of Argentina, on the south side of the mouth of the River Plate. Our broad aim was to drive due west to Mendoza, in the foothills of the Andes about 1,300 kilometres north of San Martin, and then to drive southwards towards San Martin, searching all the time for bumblebees. No one knew how far north or east the buff-tails had spread, since there are very few entomologists in Argentina. We knew we must hit their advance guard at some point along our route. In the meantime, we wanted to record any other bumblebees that we could find. Seven bumblebee species are said to be native to Argentina in addition to *dahlbomii*, but very little is known about their distribution and less still about their ecology.

Leaving Buenos Aires turned out to be harder than expected. The only road from the airport that we could find headed due north into the city centre. At that stage we had not managed to acquire a map, so we had little to guide us and became hopelessly lost trying to find a road west in a bewildering maze of streets thronged with battered, smoking cars of ancient vintage, gleaming new 4x4s, horse-drawn carts and the occasional goat. There were no signs to help us, few road markings, and dangerously deep potholes even in some of the larger roads. The local driving style seemed to involve considerable flexibility with regard to the rules of the road, if indeed there were any, with cars weaving past us on both sides, swerving erratically to avoid potholes, domestic animals and each other. Right of way seemed to belong to the person going fastest or with the most dilapidated vehicle, which meant it never belonged to me. It was altogether a nerve-shredding experience. Fortunately, the locals did not seem to mind my hesitant and slow progress, for not once did we get hooted at despite many near-collisions. I soon completely lost my sense of direction, and asked Jess to try to work out which way was west from the position of the sun. This was rather tricky since it was about midday in mid-summer, so the sun was very nearly vertically above us, but she did her best. Somehow, after an extensive tour of the affluent tree-lined boulevards of downtown and countless circuits of the less salubrious parts of the city, we emerged on the right road sometime in mid-afternoon, and began our journey west.

As soon as we left the urban sprawl, we entered the largest soya bean field I have ever seen. Argentina is one of the world's biggest exporters of agricultural products, producing roughly one quarter of all the world's soya beans and 5 per cent of its beef. Buenos Aires is surrounded by a vast, flat, fertile plain which stretches unbroken for 800 kilometres to the west, covering roughly 500,000 square kilometres, more than twice the total area of the UK. Technically it consists of many thousands of huge rectangular fields, rather than just one, but because they are separated by the narrowest of fence lines with no hedges and very few trees, the appearance was of one vast field that went on for ever. Giant silver silos for bean storage reared up at intervals on the skyline, which was otherwise unbroken. It is an impressive example of man's ability to dominate nature, to eradicate pests and weeds and grow a single crop at the expense of all else. It was interesting to see the use of genetically modified crops being boldly advertised. In Europe, GM crops are regarded with deep suspicion, often labelled 'Frankenstein' crops in lurid tabloid headlines. Here, the roadside fences carried large placards shamelessly proclaiming the particular variety of GM crop that formed the sea of green beyond. I am told that most of these soya beans are shipped halfway around the globe to China where they are used to feed beef cattle or power biofuel plants. As I gazed over this vast expanse of uniform green, I couldn't help but reflect that it ought not to be beyond our wits to devise less environmentally damaging and more efficient ways of using our world's precious resources.

I generally love to visit new countries, largely because of the opportunity to see wildlife and landscapes I have not encountered before, but there was little to enjoy here. For a biologist this region of eastern Argentina was rather depressing – it can be no surprise that the bellicose bumblebee has disappeared, along with almost everything else that presumably used to live here. Clearly pesticides were heavily used because the few streams we crossed were devoid of fish, amphibians or the predatory birds such as herons and egrets which one might expect to see in abundance in the sub-tropics. All of the streams had been canalised into straight channels, often lined with concrete, and most were little more than running sewers, stinking and lifeless. Clearly the same crop was grown year after year on the same soil, rather than using any type of crop rotation, which meant that artificial fertilisers had to be poured on in bucketloads, which then added to the pollution of the streams, causing blooms of toxic algae. The few towns we passed through were mostly dismal, dusty places, consisting largely of farm machinery sales and repair shops and agrochemicals merchants.

I couldn't help but wonder what this land had once looked like, before man turned it over to his exclusive use. The fertile soils suggest that it may have supported a dense, subtropical forest, which presumably would have contained all sorts of interesting creatures. Of course we humans need to eat, and farmers need to make a living, but it seemed a terrible shame that there is nothing whatsoever left of what must once have been an immense forest.

To be fair, there were a few interesting highlights. On our second day driving across this endless plain we stopped for lunch in the shade of some Australian gum trees incongruously planted by the roadside, from the branches of which were suspended several huge, colonial nests of green parakeets – tremendously raucous, noisy birds which peered down at us curiously. One particular species of dragonfly seemed to have developed resistance to the pollution in the streams and had proliferated, their wings shimmering in the sunshine as they hawked across the road. It was impossible not to hit them as we drove by, and small brown birds of prey which we eventually identified as chimango caracaras sat by the roadside feeding on the ready supply of dragonfly corpses. Near the towns, dead dogs were frequent and their swollen corpses often attracted the much larger southern crested caracara – impressive and ferocious-looking Lewis Carroll caricatures of birds, eagle-sized beasts with crested heads, bald red faces and disproportionately long lemon-yellow legs. They were not alarmed in the slightest by our passing and stared at us challengingly with unblinking orange eyes.

Since we were supposed to be looking for bumblebees, we stopped every eighty kilometres or so along our route to hunt for them, doing one-hour timed searches as I had done previously in many other places, such as on Salisbury Plain and in the Gorce Mountains of Poland. We chose spots to search where there were at least a few roadside weeds, usually patches of yellow star thistles, creeping thistles and teasel, all of which are non-native plants which were presumably accidentally introduced by early settlers. Try as we might, we could find no bumblebees of any sort, although honeybees were often abundant and we occasionally found huge black carpenter bees, which I briefly got excited about in the hope that they were the black bumblebee *Bombus atratus*. These magnificent but slightly fearsome insects are the size of bumblebees, jet black with an oily purple sheen, but are actually solitary bees and despite the superficial resemblance are only distantly related to bumblebees.[22]

[22] They are named carpenter bees because they chew holes in wood to make tunnels in which to nest. Many years ago, when we had just left university, my old schoolfriend Dave (the pigeon poo thief) and I decided to cycle across the Sahara Desert to Cameroon in West Africa. It was a ridiculous venture, and not entirely successful, although we did have some great experiences and got a good part of the way there. One difficulty when camping in the desert was finding sufficient fuel for a campfire on which to cook dinner. We often resorted to burning dried camel dung, which worked quite well except that it left a horribly sticky tar on the underside of our cooking pot. On one memorable occasion, as dusk was settling in and we were setting up camp for the evening we noticed a clump of brittle, leafless shrubs from which we snapped numerous branches, sufficient to treat ourselves

Towards the end of our second day of travelling west, the land slowly began to look a little less fertile, the crops less vigorous and more patchy. My GPS told me that we had been imperceptibly climbing as we travelled, and had now reached about 2,000 feet in altitude, although we had not seen anything that could remotely be described as a hill. The traffic had become steadily less frequent, so that now there was often no other vehicle in sight. After the town of San Louis the arable belt suddenly ended, and before us stretched an endless plain of low, spiny scrub, the arrow-straight and empty road ahead the only mark that man had made upon the landscape. It was a dramatic change after the intensive farmland, although no less monotonous in appearance. We stopped and stretched our legs, wandering amongst the thorny waist-high acacias. Insects abounded; tiny native bees swarmed over the few shrubs that were in flower, and a diversity of beetles scurried along the stony ground. The silence was broken only by staccato bursts of song from cicadas, endearingly ugly, bulbous-headed insects which perched among the twiggy scrub. Black vultures circled overhead, presumably hoping that we might expire. In the distance, we could see small tornadoes marching across the plain, twisted spirals of wind lifting dust and leaves high into the sky.

We continued westwards, the lack of any landmarks or bends in the road making it sometimes feel as if we were not moving at all. Every eighty kilometres we repeated our search for

---

to a good blaze to keep off the evening chill. Sadly, in the twilight we had not noticed that the branches were hollow and packed with hibernating carpenter bees. As soon as the fire began to catch the branches began to emit an unexpected buzzing noise, shortly after which the large black bees started to try to escape. Smoking, singed and angry bees were soon hurtling in all directions, making our evening under the desert stars much less restful than we might have hoped.

bumblebees. This was now much more fun as there were plenty of new insects to see, but there were no bumblebees to be found. We holed up for the night in a small town called La Paz which unexpectedly appeared ahead of us. I couldn't imagine what its inhabitants did for a living. After La Paz the landscape became a little greener, and a few brave souls had tried to carve out fields and plant vineyards along the roadside, though none looked very productive. The road continued to climb very gently, and as it did so the vineyards became more prosperous, larger, neatly maintained, and with healthy rows of vines marching across the landscape. A blue-grey bank of cloud hung on the western horizon directly ahead. The map suggested that we should be nearing the Andes, but the landscape still appeared to be more or less featureless. As we approached the city of Mendoza, the cloud bank grew taller and began to gather texture, folds and shadows. Jess and I realised almost simultaneously that it was not a cloud bank at all, but an abrupt line of vast mountains, their tops sheathed in clouds. We stopped and spent a while gaping and taking pictures that utterly failed to capture the grandeur of the scene. At this point along their length the Andes are at their highest; Aconcagua, the highest mountain in the Americas, is just a smidge under 7,000 metres tall.

Mendoza nestles in the foothills of these mighty peaks, and is the centre of Argentina's wine-growing region, rightly famed for its juicy, rich Malbec reds. Vines thrive on the cool mountain slopes, watered by the snow melt that pours down in spring from the ice-capped mountains above. By the time we arrived in Mendoza we had driven about 1,100 kilometres and conducted about fifteen entirely fruitless one-hour searches for bumblebees. We were starting to wonder whether there were any native Argentinian bumblebees left, and if so, where they were to be found. Most bumblebees prefer cool climates, so our next logical move was to try to get high up into the Andes. Old museum records suggested that *dahlbomii* used to occur this far north, and that there should be several other species present. West from Mendoza, we found an unsurfaced track which switchbacked up into the mountains. The vegetation changed from vineyards to semi-desert rocky slopes dotted with barrel-shaped cacti, and then to more lush, sub-alpine scrub. As we slowly approached Aconcagua the sky grew darker, and lightning flickered amongst the leaden clouds which shrouded its peak. It was here that Jess finally spotted our first South American bumblebee. It was a queen of *Bombus opifex*, a sizeable, largely yellow bumblebee with a reddish bottom. She was feeding on a yellow star thistle and quite unconcerned by our presence so we sat and admired her while the thunder rumbled overhead. Hunting around amongst the shrubs, we quickly found more *opifex*, mostly workers, many of them feeding on thistles of one sort or another. Later, as we followed the road up to 2,000 metres altitude, we entered an area where wild snapdragons were abundant, and these too the *opifex* seemed to love.

We spent that night in the small town of Uspallata, a pleasant, low-key tourist base for hiking in the high mountains and rafting on the tumbling rivers. We celebrated with a bottle of 'Andes' beer – in the circumstances it seemed rude not to – followed by a delicious steak each and a bottle of local red wine.[23] It was surprisingly warm given the altitude and we sat outside to eat, watching iridescent green hummingbirds darting among the purple

[23] Argentinian cuisine seemed to involve almost no vegetables – huge steaks and chips being the standard fare. Breakfast in the various hostels and hotels we stayed in invariably consisted of sugary croissant-like pastries known as *medialuna*, slices of cake, and a peculiarly grim and unnaturally brightly coloured orange squash. I felt in danger of going down with beriberi or scurvy by the end of the trip.

blossoms on the trees lining the street while Aconcagua brooded above us. It was a thoroughly pleasant evening.

The next morning we headed further into the mountains, to the border with Chile at nearly 3,000 metres altitude. We were hoping that we might find populations of *dahlbomii* this far north in the cool of the high mountains, but search as we might we found none. We did find a few more *opifex* at intervals, many of them feeding on a pretty pink daisy which grew in dense, silver-leaved clumps. Storm clouds plagued us all day, often catching us out in the open with our nets, drenching us in huge icy drops of rain. Towards the end of the day we headed back to Mendoza, and from there began the southward leg of our 6,000-kilometre journey, towards San Martin. Over the next few days we followed the main route south, which ran along the flatter land to the east of the Andes past affluent bodegas and lush fields of vines and apple orchards. Wherever the opportunity presented itself we detoured west back up into the mountains to search for bees, subjecting our poor hire car to a thorough battering on the rough tracks. In large part the mountains were heavily grazed by vast herds of goats, tended by gauchos on horseback and wearing traditional costume: broad-brimmed hats, ponchos and spectacularly voluminous trousers. I remembered learning about gauchos in geography lessons at school – I spent hours drawing one, as I recall – but I'm sure we were taught that they herded something more manly than goats. We pondered whether they had dressed up especially for the tourists, but this seemed unlikely given that we were apparently the only visitors to these remote mountain valleys. The goats had clearly eaten most of the flowers and we found few bees, although there were a few isolated populations of *opifex* as far south as the rural market town of Malargüe, some 300 kilometres from Mendoza.

After Malargüe, the route left the vineyards behind and wound through rocky desert for nearly 160 kilometres before coming to the crossing of the Rio Grande. The broad, fast-flowing brown river emerged from a steep-sided rocky ravine to the west, spilling into a flat-bottomed valley where it braided into numerous channels between banks of gravel. On the far side of the bridge lay a dead horse with what appeared to be a fence post protruding from its abdomen. Just beyond the ill-fated beast was a small cluster of half a dozen simple houses amongst a few acres of verdant green fields which stood out in the otherwise arid brown landscape. It appeared to be a likely spot for bees, so we stopped to search. The fields were well defended behind formidable fences made by weaving together the thorny branches of acacias – perhaps to keep pumas from the livestock within. They were also effective at excluding entomologists, so I skirted around the little patch of green fields along the shingle of the river bed. Peeking through the fence at a little patch of overgrown pasture, I noticed flowers of a distinctive bluish colouration – it was unmistakeably viper's bugloss. If there were any bumblebees here they would surely be on this, as it is a flower much-beloved by bees of all sorts for it produces copious quantities of sugar-rich nectar. In my mind, viper's bugloss is intimately associated with happy summer days hunting for bees, for it is found in most of the very best places for bees in Britain: Salisbury Plain, the shingle and dunes of Dungeness and the coastal dunes and marshes of the Thames Estuary. It is an invasive weed in New Zealand, in the drier parts of which it has taken over large areas of sheep pasture, covering the landscape in a blanket of cobalt-blue flowers and forming a mainstay of the diet of the British bumblebees that now live there.[24] I always have a

---

[24] In *A Sting in the Tale* I describe how bumblebees were introduced to New Zealand from Kent in 1885, and attempts to reintroduce the

patch in my garden, for it is easy to grow and looks magnificent in the herbaceous border. It is even said to be good for treating snake bites (presumably the origin of the name) and bee stings, though I have never tested this.

Luckily, there was a point where I could clamber over the thorny fence, so I did so a little sheepishly, expecting an irate farmer to shout at me at any moment. There were swarms of bees – carpenter bees, honeybees, all manner of solitary bees, and bumblebees. A moment's inspection revealed that these were the buff-tail invaders – not just one or two, but dozens of them, everywhere, feasting on the viper's bugloss. We had met their vanguard, advancing north from San Martin.

In some ways it was lovely to see such familiar creatures. Buff-tails are by far the easiest bumblebee to breed in captivity, so I have spent many happy hours studying them in various experiments over the years. But although we knew we must meet them somewhere along our route, I had hoped that it would not be quite so soon. A quick look on the map confirmed that we were still about 900 kilometres north from San Martin, a distance which the buff-tails had covered in just eight years. Given that they normally have only one generation each year, and that the nests themselves clearly cannot move (unless they are cleverer than I suppose), all of this ground must have been covered by queens before they built their nest each year, each dispersing on average just over 100 kilometres. This is extraordinary – in Tasmania and New Zealand, the buff-tails spread much less rapidly following their introduction, covering just a few kilometres each year, although they did eventually colonise the entirety of both islands. We had had no idea that queen

---

short-haired bumblebee, a species which has since gone extinct in the UK, back from New Zealand to its last known British haunt at Dungeness.

bumblebees were even capable of such epic flights. Although these were an alien species, potentially doing terrible harm to the indigenous bumblebees, I couldn't help but feel affection and not a little admiration for these hardy little creatures, such a familiar sight in Europe, forging a successful life for themselves in this new world. It was, after all, not their idea to come to South America in the first place, and we can hardly blame them for making a success of it.

I called Jess over, and between us we caught and pickled a sample of bees so that we would be able to screen them and determine which diseases they were carrying. The bumblebees were so numerous that at one point I accidentally caught two in my net at once, a schoolboy error. Getting a bee from a net to a pot is fairly straightforward with a little practice, but fraught with difficulty if there is a second angry bee in the net at the same time. A sensible person lets them both go, but I can never resist the challenge of trying to get them both into a pot simultaneously. The outcome is almost inevitably a stung hand, and so it was on this occasion. Fortunately, Jess proved to be particularly adept with a net and caught plenty while I was floundering around, cursing and sucking my throbbing thumb.

We set off southwards once more, following the Rio Grande for many miles. Despite the water there were few flowers and no bees. We entered a spectacular area of active volcanoes, with several neat, conical peaks a little to our west. The river had carved its way through jet-black, solidified lava flows, and our road deteriorated to a rough gravel track as it attempted to follow the river through this magnificent but inhospitable landscape. It was baking hot and the black rocks soaked up the sun's heat, creating a shimmering heat haze. There was almost no plant life, indeed almost no life of any kind apart from house flies, which descended on us in droves the moment we stopped to eat our picnic lunch. I had

had a similar experience when sitting in the middle of the Sahara thirty years earlier. Quite how these creatures manage to sustain themselves in such places is utterly beyond me.

After 250 kilometres or so we emerged into a welcome patch of greenery surrounding the small town of Chos Malal, and here we immediately found buff-tailed bumblebees in abundance. I couldn't help but wonder how the buff-tails had managed to cross the landscape we had just come across. It had been mildly arduous for us, with the benefit of an air-conditioned car, lots of bottled water and a nice packed lunch – how a queen bee could find her way across that flowerless wasteland was hard to conceive.

From here on, southwards to San Martin, the story remained much the same. The road travelled through several more expanses of rocky, arid desert, but wherever there was a patch of greenery there were numerous buff-tailed bumblebees, and no matter how hard we searched there were no native bumblebees to be found, and in particular, there were no giant golden ones. We would catch and pickle a sample of buff-tails, then move on. As we drove we pondered why the buff-tails had become so common here. They presumably benefited from the plentiful European weeds, such as the viper's bugloss and thistles, that we saw growing along the roadsides. Perhaps they also enjoyed freedom from competition with the many other short-tongued bumblebee species found in their native European range, such as in Poland. It seemed likely that they would have left behind many of their natural parasites, creatures such as the tracheal mites that infest their airways and conopid flies that eat them alive from the inside. Of course once we were back in the UK our samples of bees would allow us to find out exactly what parasites and diseases they did have.

Some eighty kilometres before San Martin the horizon ahead of us acquired a murky, grey appearance. It looked a little like the dense smog I have seen above cities such as Los Angeles and

Seoul. The road, which had been taking us across a rocky plateau, suddenly plunged over the lip of a mighty ravine, and zigzagged perilously down the cliff face on narrow ledges. I pulled over. It should have been a most splendid view, a slightly less grand version of the Grand Canyon, except that the air was thick with what I took to be smoke, and we could only occasionally see across the valley as the smoke ebbed and gusted in the brisk wind. We got out, and it was immediately clear that it was not smoke but grit and dust, which stung our eyes and made us both sneeze. Huge black birds soared on the updraft, eight of them in total. It was hard to see them well, squinting into the gritty wind, but I noticed that they had white ruffs around their necks and that they were further away than I had at first realised, and hence they were larger – they were in fact condors, the largest flying birds on earth. It was a spectacular and eerie sight, these monstrously large vultures hanging silently in the swirling dusty air.

We watched them for as long as our eyes could bear, and then headed down to the canyon floor. Soon afterwards we passed through the small town of San Junin and the vegetation changed dramatically. Lush green forests appeared in the valley bottom, the first forests we had seen in a journey of roughly 2,500 kilometres, and as the road wound westwards we emerged into a rolling landscape of forested hills, clear babbling streams and icy blue lakes. Were it not for the dust which still hung in the air, it would have been exceptionally beautiful. These forests stretch south to San Martin and San Carlos de Bariloche beyond. Most of Argentina lies in the rain shadow of the Andes – the prevailing winds come from the west and drop their rain on Chile as they rise over the Andes, so that the air is dry by the time it sweeps down the eastern side of the mountains. However, in this area the mountain chain is lower, allowing moist air from the Pacific to reach this far east and support these lovely forests. This, of course, is also why this is the area where introduced bumblebees, both ruderal and buff-tails, first made the crossing from Chile to Argentina.

We stopped once more to search for bees. There were plentiful flowers, and buff-tails were abundant. Spectacular swathes of orange-flowered alstroemeria and purple fuchsia grew among the trees. I caught the first and only ruderal bumblebee that we saw on the entire trip, a queen feeding on a dusty hollyhock flower by the road-side. Until 2006 the giant *dahlbomii* had been a common sight in this area but, as we had been led to believe, there were none to be found.

As we travelled south-west the dust became thicker, and the bees less frequent. The ground became covered in a layer of it, the vegetation smothered under a blanket of pale grey, and it finally dawned on me that this was ash. We had heard that a volcano had erupted just over the border in Chile some two months earlier, and that it was continuing to throw out pumice and ash, but coming from a land where such things are unknown, it had taken me a while to recognise what I was looking at. We passed lakes which were partially covered in layers of pumice at the leeward end – pumice floats, and we later learned that, just after the first and strongest eruption, it was possible to walk across these lakes, so thick was the floating layer. The few buff-tailed bees we saw looked pretty sad; they were themselves covered in dust, and were trying to forage on ash-covered flowers. Their pollen loads must have been mixed with ash, and I suspected that this may have killed the developing grubs, or else perhaps many of the nests were simply buried under ash and the bees unable to escape. Certainly bees of any type were few and far between.

We passed through San Martin, which proved to be a pretty tourist town on the edge of a long lake, popular for hiking in summer and skiing in winter. The houses had mostly been built by Swiss and German settlers out of timber and styled upon Alpine chalets, making the area seem like a slice of Switzerland plonked down in South America, an impression heightened by the surrounding snow-capped mountains. We stopped for a drink in a café that sold nothing but twenty-five variants of hot chocolate, plus huge slices of delicious chocolate cake to accompany it.

In Bariloche, a day's drive beyond San Martin through beautiful forested scenery during which we left most of the ash cloud behind, we met up with Carolina and her PhD student Marina Arbetman. Marina had recently started her PhD on the impacts of the invading buff-tailed bumblebees. She had been trying to chart their spread southwards from San Martin, and had recorded them as far away as El Calafate, some 1,200 kilometres to the south. She had seen almost no *dahlbomii* anywhere within the expanding range of the buff-tails, but a colleague had reported seeing one a few days earlier by a nearby lake. Marina took us to look for them. There was much less ash here and the area was stunning, with jagged-toothed snow-capped mountains framing the glorious turquoise lakes. The area is protected as a national park, swathed in dense forests, and rich in wildflowers. We walked along the shore of Lago Nahuel Huapi, at the place where the *dahlbomii* was recently seen and where Carolina has been regularly counting bumblebees for many years. The shore was lined with Arrayan trees (*Luma apiculata*), gnarled and twisted by the fierce winds which must blow across the lake in winter. The trees were covered in delicate white blossom, flowers formerly favoured by *dahlbomii*, but all we saw were buff-tails. There were plenty of other things to see. We spotted pigeon-sized blue, white and rust-coloured ringed kingfishers hunting by the lake shore, and flocks of noisy olive-green parakeets flew past. Nonetheless, it was hard not to be a little depressed. From Mendoza southwards for 1,300 kilometres, we had been within the historical range of *dahlbomii*, but we had not seen a single one. Now it was clear that they were close to extinction throughout almost all of their known range. Beyond El Calafate, it is just another 300 kilometres to Tierra del Fuego, and then about the same distance again to the very tip of South America. At the current rate of spread, it seemed likely that buff-tails would reach the tip of South America within two to three years, and that might be the end for *dahlbomii*.

So what is causing the decline of the giant golden bumblebee? We discussed this at length with Marina and Carolina. It seems unlikely to be competition from buff-tails. The difference in their tongue lengths means that they have different flower preferences, although they do feed on some of the same flower species. Long-tongued and short-tongued species readily coexist elsewhere in the world, and in places such as Poland dozens of species happily live alongside one another. Also, the speed of decline of *dahlbomii* seemed to be far too fast to be explained by competition.

We discussed the possibility that it might be due to disease. Little is known about bumblebee diseases in most parts of the world, but we do know that they can suffer from a diverse range of infections caused variously by protozoans, bacteria, fungi and viruses. When the Spanish conquered the new world they accidentally infected the native Americans with a host of European diseases to which they had little or no resistance. Diseases such as measles, from which Europeans usually recover swiftly and without any permanent ill-effects, proved to be lethal. The impact was devastating, with perhaps as much as 95 per cent of the entire population of the Americas wiped out over a few decades. Entire civilisations, such as that of the Aztecs, collapsed into chaos and then were easily overrun by a tiny force of European invaders. Recent evidence even suggests that substantial parts of the Amazon basin, long thought to be primary, undisturbed rainforest, actually supported large civilisations and extensive areas of farmland until about 500 years ago when they were wiped out by these diseases, which spread far ahead of the Europeans themselves. The process was largely one way, but the American natives did give the Europeans syphilis in small revenge for their fate.[25]

Was something similar going on in bumblebees? My PhD student Pete Graystock had recently performed some ingenious experiments in which he had shown that bee diseases can be transmitted between bees of different species without the bees actually ever needing to meet. Every time an infected bee visits a flower she contaminates it with a sprinkling of parasites, perhaps contaminating the nectar with her mouthparts, or the petals by contact with her feet and body. The next bee that comes along may unknowingly pick them up, and either ingest them or carry them back to her colony. Pete came up with the neat term 'florally transmitted diseases', FTDs for short. If buff-tails were carrying something to which they were largely resistant, it could easily spread to native bees. However, testing this hypothesis is difficult. There are virtually no *dahlbomii* left, so we can't study them to see what is killing them. If one were lucky enough to catch one of the few survivors, it would presumably be one of the few that had escaped the disease, or perhaps one of a small number with natural resistance. Studying the invaders can tell us what diseases they have – this was why we had been catching samples of buff-tails – but it cannot tell us which disease is killing the natives, if indeed it is a disease.

[25] There is some debate about this, with some contending that syphilis occurred in Europe before the rediscovery of the Americas, but the first clear account of the disease in Europe is from 1494 in Naples, appearing suspiciously hot on the heels of the first European explorers returning from the Americas.

Marina had already started work on this, and had found that many buff-tails near Bariloche were infected with *Apicystis bombi*, a disease known from bumblebees in Europe and North America. In Europe this disease is sometimes fatal to bumblebees, but many individuals seem to carry the pathogen without showing symptoms. This disease had not been recorded previously in South America, but on the other hand hardly anybody had ever looked so it is hard to know what this means. It is quite plausible that this disease arrived with the buff-tailed bumblebees and is the cause of the *dahlbomii* decline, but we cannot say for sure.

We subsequently studied the pickled buff-tails that we collected in Argentina and found many of them also to be infected with a second disease, *Crithidia bombi*. This is a trypanosome, a relative of the sleeping sickness spread by tsetse flies in Africa, which causes gut infection in bumblebees. Infected bees suffer a range of symptoms – their ovaries tend to be smaller, and they seem to be a little more stupid, being less well able to learn and remember which flowers are most rewarding. *Crithidia bombi* had been found in South American bees before buff-tails were introduced, but the invaders were carrying genetically distinct strains that presumably came with them from Europe. Maybe this is the cause of the collapse of *dahlbomii*?

Clearly more work is needed to find out exactly what is going on. It should be possible to use museum specimens of *dahlbomii* to see what diseases they used to suffer from before buff-tails arrived. It might sound tricky to tell what diseases a long-dead bumblebee on a pin might have been carrying, but modern genetic techniques are very sensitive and ought to be able to detect the DNA of the pathogen. Marina has already had a go at this, and she has not detected any *Apicystis* in a small sample of old *dahlbomii*, but she did not have many specimens to play with and she could not be sure that the negative result was genuine or whether

her technique was faulty. What she really needed were old museum specimens of some bumblebees which we knew had been suffering from *Apicystis* when they were killed as 'positive controls', but no such bees are available.

The ruderal bumblebees may also offer a clue as to what is going on. Ruderal bumblebees in Argentina appear to have been almost as badly affected by the arrival of the buff-tails as have *dahlbomii*. This is intriguing since the ruderals also originate from Europe, so one might expect them to be resistant to European diseases. On the other hand, those in Chile and Argentina were brought from New Zealand, where they had been living for about a hundred years since their original introduction from England. If they had not been exposed to a particular disease for a hundred generations in New Zealand they might well have lost their resistance to it. Thus the cause of the problems, if indeed it is a disease at all, must be something which occurs in Europe but was not transferred to New Zealand when bumblebees were introduced there. As luck would have it, I had collected samples of ruderal bumblebees from New Zealand some years earlier so it would be easy enough to check what diseases they have there – I just need a little funding from somewhere to cover the costs.

With a little more detective work, I hope that we can diagnose the cause of the demise of *dahlbomii*, but what then? We cannot vaccinate the remaining bees in Tierra del Fuego before they are wiped out. It might conceivably be possible to set up a captive breeding programme, but this would be extremely tricky, for they have never been bred in captivity before, and in any case we would probably never be able to release them back into the wild unless future technology can provide some permanent solution to the problem. One of the more innovative suggestions that has been put to me is that we should evacuate some *dahlbomii* from Tierra del Fuego to the Falkland Isles, which have no native bumblebees at all. The climate there is probably quite similar, though I have no idea whether there are suitable flowers, or whether *dahlbomii* would survive. Perhaps an Anglo-Argentinian project to save a bumblebee could help to repair the poor political relations between our two countries? On the other hand, might we do more harm than good by going down this route? There are no doubt native pollinators of some sort in the Falklands, and they might suffer from competition with a new bumblebee, or even be infected by diseases that naturally occur in *dahlbomii*. All things considered, and appealing though this idea is, I suspect that we should learn our lesson and not meddle even more with things we do not well understand, for fear of making things even worse.

Potentially, there are more dire consequences of the introduction of buff-tailed bumblebees than the demise of *dahlbomii*. We have no idea how the other South American bumblebees will respond when faced with this new invader. If the disease hypothesis is correct, then it seems probable that the other native species will also be susceptible. The lovely *opifex* were just a short hop north of the advance wave of buff-tails heading for Mendoza in 2012, and at the time of writing this in 2015 it seems highly likely that the populations Jess and I found have been overrun. Will they too disappear? Further north still, there are many more bumblebees. South America has twenty-four in total, many living in the Andes of Peru and Colombia. How far can the buff-tails go? We do not know. It should be possible to predict this from their known climatic tolerances – in Europe they range from Scotland to the south of Spain, and also to Morocco, Tenerife and eastwards into the arid lands of the Middle East. They are clearly highly robust and adaptable creatures, perhaps more so than any other bumblebee, and with the cool mountain chain of the Andes to follow northwards, they may be able to penetrate much of South America. Only time will tell.

It is not entirely without hope. Tierra del Fuego may prove to be too cold for buff-tails, although I doubt it, but if this were so then it could remain as a refuge for *dahlbomii*. There may be undiscovered, remote mountain populations of *dahlbomii* in parts of the Andes where buff-tails cannot penetrate. The occasional sightings of *dahlbomii* near Bariloche, eight years after buff-tails arrived there, provide perhaps the brightest prospect. Just as a small proportion of the Native Americans survived the onslaught of disease, so it may be that these surviving *dahlbomii* possess a degree of resistance. If enough have survived, perhaps their population may slowly recover, for the habitat in which they once thrived is mostly still there, waiting for them.

After Bariloche, we headed east. I would have loved to continue southwards far enough to finally see some *dahlbomii*, but we didn't have time, and we knew what we would find in that direction, for Marina already knew how far buff-tails had spread south from Bariloche. Instead, we turned east, away from the Andes, to complete the picture of how far buff-tails had spread, and to find out whether there were any native bumblebees between Bariloche and the Atlantic coast. We soon left the verdant forests behind and were once more in an arid, inhospitable, rocky landscape. We drove past fantastic folds of limestone, layers of rock once laid down in a seabed perhaps 100 million years ago, now tilted on their side and buckled by volcanic activity. To make a change from bee-hunting, we spent a while clattering around amongst the rocks looking for fossils. They teemed with ammonites, long-extinct molluscs from the ancient sea. There have been some remarkable fossil finds in Patagonia, including several new species of dinosaur, but we didn't find anything quite so dramatic.

Although we didn't find any dinosaur fossils, the following day we did round a corner to see a real dinosaur eating a man, or so it at first seemed. For reasons that remain obscure, someone had constructed a life-sized fibreglass replica of one of the larger predatory dinosaurs and placed it in the middle of nowhere. A passer-by had chosen to park his car under the gaping jaws, climb onto the roof and then jump up to hang from the teeth, just as we came into view.

As we travelled towards the distant east coast, we searched every good patch of flowers for bumblebees, but we were not to find a single one of any type on the rest of our trip. We saw many other wonderful sights – including rheas, giant flightless birds first described in detail by Charles Darwin in his travels on the *Beagle*. Most were family groups trailing a dozen or more youngsters. There were guanacos and llamas, trotting in small groups across the dry plains, but no matter how hard we looked we could find no bumblebees. Buff-tails seem not to have penetrated far to the east, which is peculiar given the extent of their spread north and south, and also that the prevailing wind would carry them eastwards. There is an extensive area of near-desert immediately east of Bariloche, but no worse than those they must have crossed to move north. Three hundred kilometres to the east we found well-watered, flower-rich areas where I would guess that they could thrive, and I suspect they will arrive there eventually.

Further east we crossed the Pampas, a vast area of grassland dotted with acacia trees, which in spring has many flowers. It seemed as if it ought to be a good place for bumblebees of some sort or other, but we found none. Here the gauchos did have more impressive beasts to herd, for this region supports fifty million head of cattle. We had already eaten plenty of beef, but that evening we carelessly ordered a 'mixed barbecue', expecting more delicious steak cuts. This was perhaps naive, for we were served an interesting selection of body parts which we struggled to identify, but which almost certainly included slices of barbecued penis and an awful lot of tripe. Eventually, after covering nearly 6,000 kilometres of bumpy and dusty roads, we emerged on the coast near the city of Bahia Blanca. There was a pleasant sandy beach so we rushed in for a refreshing dip, but I was almost immediately stung by a jellyfish, which rather took the fun out of it. The tentacles stuck to the flesh of my feet, causing red weals to rise up as if I had been whipped.

The introduction to South America of European bumblebees is yet another cock-up in a long history of predictable catastrophes orchestrated by man. It occurred long after we should have known better. Presumably a few farmers thought they might benefit financially from having buff-tailed bumblebees in their fields in Chile, and the wildlife of an entire continent is now paying the price. It is conceivable that twenty-four species of bumblebee could be wiped from existence by this one thoughtless act. There does not appear to be any solution to the problem, no way of undoing what has happened. We can hopefully identify the agent behind the decline of *dahlbomii*, and perhaps this will be useful for informing any future debates as to the advisability of introducing non-native bumblebees elsewhere. Perhaps the native bumblebees in northern South America have some bumblebee equivalent of syphilis which might halt the spread of buff-tails.

The latest news from South America, as of January 2016, is that the buff-tails have reached the northern shores of the Strait of Magellan. At the narrowest point, the Strait is only about ten kilometres across, so this is unlikely to hold them up for long. A Chilean entomologist named Jose Montalva has successfully launched a campaign to raise the profile of *dahlbomii*, and to encourage people to send in records of sightings of both the native and invading species. A few *dahlbomii* continue to be seen in the invaded area, giving hope that these may be individuals with resistance to the European diseases. It will be fascinating to see what happens, and all we can really do now is cross our fingers and hope for the best. I also hope that one day I will get to see a giant golden bumblebee, alive and well in its native forests, but unless I can find a pressing need to go to Tierra del Fuego sometime very soon, the odds are probably against me.

CHAPTER FIVE

California and Franklin's Bumblebee

Biological diversity is messy. It walks, it crawls, it swims, it swoops, it buzzes. But extinction is silent, and it has no voice other than our own.

Paul Hawken

In spring 2013 I was invited to Davis, part of the University of California, to give a couple of talks on my research on bumblebees and to meet with the research group of Neal Williams. Neal is a bee biologist, a gangly, laid-back and exceedingly likeable guy whom I had met a couple of times at conferences, and whom I knew to be doing some exciting work on how best to boost pollinator populations in farmland in California, so I was keen to go and learn more about what they were up to. I also had an ulterior motive, for Davis is within striking distance of the last known haunts of Franklin's bumblebee.

The tale of Franklin's bumblebee is a sad one. Were it not for the work of one man, Robbin Thorp, we would know almost nothing about this species. Robbin is also based at Davis, where he worked for all of his long career. For many years he was mainly focused on honeybees and crop pollination, but in his later years he became particularly interested in wild bees, including both bumblebees and the numerous solitary bee species to be found in California. When he retired, now twenty years ago, he had the freedom to do whatever he liked and he focused his interests more on bumblebees, conducting regular field trips all over the USA and elsewhere to study them. It was Robbin who discovered most of what we now know about Franklin's bumblebee.

Franklin's is an attractive, medium-sized bumblebee, almost entirely velvet black with a smart, broad yellow collar, a little military in appearance. It was named in 1921 after Henry J. Franklin, who wrote the first monograph of the bumblebees of North and South America in 1913 and was the first to describe the species. So far as we know from Franklin's work and museum records, Franklin's bumblebee was always restricted to quite a small geographic area, on the border between California and Oregon, the smallest known natural range of any bumblebee. As is so often the case, we have no idea why it was restricted to this particular area – it wasn't extraordinarily fussy in any way that we are aware of, but something kept it tied to this region. It fed on common enough plants, including lupins, Californian poppy and horsemint, amongst many others. We don't know much about where it nested, but it probably used old rodent burrows like many other bumblebees. Until about 1994 it was quite common if you knew where to look, which Robbin did.

Tragically, in the mid-1990s something awful unfolded in North America. A whole group of bumblebees very rapidly disappeared from almost the entire continent, in just a couple of seasons. The bees that went were all close relatives, belonging to the subgenus *Bombus* (to clarify, all bumblebees belong to the genus *Bombus*, but this is divided up into lots of subgenera of which one is also, confusingly, called *Bombus*). This group included some of the commonest bumblebees in North America: the rusty patched bumblebee, the yellow-banded bumblebee, and the western bumblebee. Suddenly, these species went from plentiful to scarce or absent. They disappeared entirely from vast tracts of their range.

Luckily, most of them have hung on somewhere; for example, the rusty patched bumblebee can still be found at a few sites in Illinois and Iowa, in the north-west of its originally very large range, which once stretched across eastern North America. The yellow-banded bumblebee seems to be doing okay in parts of Maine and Vermont, while the western bumblebee is still reasonably common in Alaska, though it has vanished from perhaps three-quarters of its range. Sadly, Franklin's bumblebee is the exception. Franklin's also belongs to this subgenus, but unlike the others it was already a local bee, with a relatively tiny range spanning an area of perhaps 300 by 100 kilometres. In 1995 Robbin noticed that there were fewer than usual, and year on year the population plummeted. By 2006 they were gone. Ever since, Robbin has returned to their former haunts near the California–Oregon border to hunt for them, but without luck.

So what happened? Our best guess is something similar to the tragedy that is unfolding in South America, although this time there was no deliberate introduction of foreign bumblebees. The cause of the demise of this whole group of bees may have its roots in Belgium in the 1980s, with the commercialisation of bumblebee rearing. A veterinarian and amateur bee enthusiast named Dr Roland De Jonghe discovered that bumblebees provided superbly effective pollination for tomatoes, something that honey-bees are pretty hopeless at, and he began rearing buff-tailed bumblebee nests for sale. Prior to this, scientists had reared small numbers of nests for research purposes, but nobody had tried to gear the process up for mass production. Demand for De Jonghe's nests was huge,[26] not just within the Netherlands but also from abroad, and by 1990 several factories had sprung up in Belgium

---

[26] Globally, we produce a mind-boggling two trillion tomatoes every year, most of them pollinated by bumblebees.

and Holland with rival companies vying to satisfy the global thirst for bumblebee nests.

North American tomato growers wanted access to this new technology, but unlike in Chile, their regulations prevented importation of alien species. Techniques for mass-rearing North American species had not been developed at the time; exactly how the European bumblebee breeders do this is shrouded in secrecy. I was once lucky enough to be shown round the three biggest European factories, but each company insisted that I sign a non-disclosure agreement so that I couldn't reveal their secrets. What was intriguing was that each factory has developed quite different rearing techniques, but because they will not share knowledge with their competitors they are unable to optimise the process. Sadly I cannot tell you any more or I might find myself in court.

Clearly in this climate of secrecy folk in North America could not simply ask how to set up their own bumblebee-rearing factories. The story as to exactly what happened next is hazy, but legend has it that queens of some North American species were sent to the European factories to see if they could persuade them to found nests, and that these nests were then returned to North America. It sounds like an entirely plausible scenario, but it is hard to find documented evidence, and the bee-rearing companies are not keen to admit involvement for reasons that will become apparent.

Unfortunately, it seems to have thus far proved impossible to mass-rear bumblebee nests that can be guaranteed to be free from diseases. Bumblebees naturally suffer from a range of diseases, including viruses, bacteria and fungi, and these are inevitably brought in to the factories when they are starting up new cultures with wild-caught queens. The bumblebee nests also have to be reared on pollen that is taken from honeybee hives, for there is no other cost-effective way of gathering the thousands of kilos of pollen needed (try collecting pollen from flowers yourself, and you'll soon realise that bees are much better at this than people). Many of the diseases of honeybees and bumblebees are shared, so this pollen provides another route for accidental introduction of bee diseases into the factories. The factories do their best to stamp out diseases as they are clearly not good for business, but the nests leaving the factories are still commonly infected. A recent study published in 2012 by an Irish research group showed that nests from all of the main European factories are routinely infected with a whole range of unpleasant pathogens.[27]

It seems very likely that the North American bumblebee nests that were returned from Europe had picked up one or more European diseases, and that these then escaped into the wild bumblebee populations in North America. Unlike the situation in Patagonia, the disease is not being spread by an invading bee species, but presumably spread through the native species. Robbin thinks it was probably a virulent strain of a fungal disease called *Nosema bombi*, causing a sort of bee diarrhoea. Whatever it was, if indeed it was a disease, we have no idea why the impact was so great on one subgenus and not on the other North American species, although it may be significant that the main species reared in the European factories, the buff-tailed bumblebee, also belongs to the subgenus *Bombus bombus*. Perhaps, also, the bumblebee species that have not been affected sometimes carry the disease without showing symptoms.

Conclusive proof for this explanation is conspicuously absent. Just as with *dahlbomii*, the collapse of these North American bumblebee populations was so fast that it caught scientists napping,

---

[27] Tom Murray, who led this work, subsequently received threats of legal action from the commercial bumblebee rearers if he did not withdraw his paper, but fortunately he did not and the threats turned out to be hollow.

and by the time anyone realised what was happening it was largely all over. Nobody could find the bodies and identify what had killed them. Those populations that hung on were presumably those that had a little natural resistance, so even if the pathogen were identified and they were experimentally exposed to it they would perhaps not die.

One clever way of getting a handle on this is to try what Marina has been doing in Argentina – use museum specimens, in this case from before 1990, and identify what pathogens they contain using genetic techniques to amplify pathogen DNA. This can then be compared to the pathogens found in the wild in North America today; anything that arrived with commercial bees should not have been present before 1990. Sydney Cameron's lab at the University of Illinois have such work underway, so hopefully we may understand more about this within a year or two, though of course that would not provide us with any mechanism for solving the problem. Once a non-native species – be it a bee, a disease or a cane toad – has been released it is rarely possible to eradicate it, and this would certainly be impossible for a microscopic parasite.

So it was that I flew out to California in late April, on an interminable dog-leg journey from Glasgow to Amsterdam to Portland, Oregon, and finally south to Sacramento. It was a beautiful spring afternoon as we flew on the final leg of the journey, and although I was tired I had my nose pressed against the glass of the window. Portland looked like a green and pleasant city, bisected by the Columbia River, and surrounded by lakes and forests. As we headed south we quickly left civilisation behind and flew over wild and remote mountainous country, dense forests from which jagged snow-capped peaks rose. Even in late April the higher lakes were still frozen and carpeted with snow, etched white amongst the endless dark forests, which I imagined to teem with bears, moose and elk. As we continued south the land below dropped and became a little less rugged, with more settlements and more pale green pastures where the forests had been cleared. This, right beneath me, was the heart of Franklin's country, centred on the rural town of Ashland in southern Oregon. Perhaps, some-where amongst the myriad green valleys and forested peaks, a few surviving Franklin's queens might be thinking about burrowing up from their hibernation chambers to meet the spring sunshine.

From Ashland, we passed over more wild, densely forested mountains, and over the 3,000-metre snowy volcanic peak of Mount Shasta in northern California. From there southwards the land dropped – the forests started to break up with homesteads visible, dotted amongst the trees, and patches of meadows, just as there were near Ashland. Then, suddenly, the forests ended abruptly and we were over a seemingly endless extent of pancake-flat arable land – the Central Valley of California, an 800-kilometre-long strip of flat, fertile land which runs due north to south, sandwiched between the Rocky Mountains to the east and the Californian coastal mountain ranges in the west. This is one of the most in-tensively farmed areas on Earth; the fields below resembled a chequerboard of greens and brown, most of them perfectly square with precisely aligned north–south and east–west boundaries with just the thinnest of fence lines to separate them. As the plane started to drop I could see that some of these huge fields were planted with regular rows of perfectly spaced fruit trees, but apart from these there appeared to be almost no trees, and precious little else in the way of natural habitat where wildlife might live – it looked rather bleak and depressing.

Davis is towards the northern end of the Central Valley, a few kilometres west of Sacramento. Despite my unfavourable first impressions of the Central Valley when coming in to land, Davis turned out to be charming – a sleepy university town, a world of sunshine, flip-flops, shady avenues, blue sky, pavement cafés and

Frisbee games in the park. It seemed a million miles from Scotland as I explored on foot the following morning. The wide, tree-lined avenues are laid out on the usual regular grid but unlike most American towns, there were hardly any cars; nearly everybody seemed to be on bicycles. The few cars on the streets were driven very slowly: in Davis the pedestrians have right of way to cross at any time. As a result, I found exploring Davis on foot slightly trying; every time I paused on the pavement to get my bearings the cars would stop, thinking I wished to cross the street. If I hesitated, they would beckon me across, smiling encouragingly, and I would feel obliged to do so. I spent much of my time politely crossing backwards and forwards over the quiet, leafy streets.

I spent the next few days in Davis, giving my talks, chatting with Neal's students, and finding out more about their work. Much of Neal's efforts are focused on understanding how bees of different sorts use the landscape – where they nest, how close those nest sites need to be to crops for them to reach and pollinate them, and what other flowers aside from those of the crops they need to keep themselves going. In many ways it was similar to what my own research group had been doing for many years, but in a nicer climate and with different types of bee. There are a huge diversity of crops that need pollinating in the Central Valley – melons, apples, strawberries, peaches, nectarines, almonds, watermelons and many more. Yet the area is pretty inhospitable to bees, for it has been more or less entirely denuded of native vegetation, so there are few places for them to nest and few wild-flowers for them to feed on. One solution is for farmers to bring in lots of honeybee hives when a crop flowers – this is what the almond growers do – but it isn't ideal. Honeybee hives are expensive to hire, and they are becoming increasingly hard to come by as honeybee colonies have been dying at unusually high rates in recent years. If the honeybee supply were to fail for some reason, such as an outbreak of disease, then the farmers would be in big trouble. In any case, honeybees aren't particularly good at pollinating some crops, such as tomatoes and strawberries, so relying entirely on domesticated honeybees is clearly not the answer.

As a result of the high demand for pollinators and the shortage of bees in this impoverished landscape, California has become something of a hotbed for bee research. Mirroring Robbin Thorp's career, for many years this work focused exclusively on honeybees – how best to breed them and provide a ready supply for wherever they were needed in California. In recent years, interest has turned to the benefits that other, wild pollinators can bring, as realisation has dawned that many of the other 4,000 or so species of bee found in North America are also useful pollinators. Some of Neal's earlier work, carried out with Claire Kremen from the University of California in Berkeley, showed that farms near patches of wild habitat – which largely means farms near the edge of the Central Valley – benefited from wild pollinators spilling out from the natural vegetation to visit and pollinate their crops. Organic farms near the edge of the valley had no need to buy in honeybees at all, whereas those that used pesticides, and those that were in the centre of the valley, had to hire in honeybees or their crop yields suffered. What was more, farms near patches of wild habitat received more reliable pollination from year to year – relying on one pollinator species is risky because that bee might have a bad year, whereas if you have lots of different ones then every year at least one of them is likely to be doing well.

Of course farmers in the middle of Central Valley can't just pick up their farm and move to the edge, and might feel envious of their valley-side neighbours, awash with bees as they are. One obvious solution to their situation might be to try to provide extra flowers and nesting sites for wild bees on the farms themselves, so that they are not reliant on there being wild areas nearby.

This is what Neal's team have been investigating – sowing strips of flower mixtures along field edges or right across the middle of fields, and planting woody hedges that can provide nest sites. The flower strips look beautiful – mixed shades of yellow and orange from sunflowers, lupins and Californian poppy, intermingled with the mauve of phacelia, and the results so far strongly suggest that, as one might have guessed, these strips do boost the wild bee population and improve crop pollination. What is more, there can be benefits that go far beyond encouraging bees. A recent field study by Richard Pywell's research group at the UK's Centre for Ecology and Hydrology revealed that setting aside 8 per cent of the crop area for flower mixes and other wildlife habitat resulted in not only an increase in the numbers of bees, but also increases in the abundance of predators of crop pest such as ground beetles, ladybirds and hoverflies. More compellingly, the yields of various crops increased steadily over the five-year period in the areas where the extra wildlife habitat was created, so that the farmer lost no yield by sacrificing 8 per cent of the land. Of course one should not extrapolate from Buckinghamshire to California, but nonetheless these kinds of studies argue powerfully that less can be more in farming. It might seem obvious that a farmer will get more yield if he plants a larger area with crops, but this clearly is not necessarily the case. Imagine if we could use these and similar studies to persuade farmers across the globe to incorporate significant patches of wildlife-friendly habitat on their farms, simultaneously reducing their need to use insecticides. Perhaps those vast crop monocultures of the Central Valley and the endless soya bean fields in Argentina would produce just as much food if there were strips of wildflowers and natural habitat spread across the plains.

While I was at Davis I had long chats with Robbin Thorp, and he showed me around Davis's dedicated bee research facility, the grandly entitled 'Harry H. Laidlaw Jr.[28] Honey Bee Research Facility', which lies at the end of 'Bee Biology Road' – I kid you not – on the outskirts of Davis. Robbin presents an imposing, Darwin-like figure – I would guess that he must be well into his seventies, tall, with a bushy white beard, albeit not yet quite as lavish as the great man's became in later life. Half of the building is still devoted to honeybee rearing and research, with experimental hives indoors in the labs, but linked to the outside world by plastic pipes through which the worker bees rush back and forth. The remainder of the building was full of eager students, either trying to identify pinned specimens of wild bees or engaged in the tricky and time-consuming business of rearing Vosnesensky bumblebee nests, which would later be used in experiments. This is one of the most common Californian bumblebees, and the nests were destined to be placed out in fields with or without wildflower strips, to see how much the flowers benefited the survival and eventual reproduction of the bee colonies.

Outside, Robbin proudly showed me their Häagen-Dazs-sponsored 'bee haven', a garden filled with mainly native Californian flowers centred around an enormous ceramic bee on a plinth. The flowers were absolutely teeming with real bees of bewildering variety. Robbin was brimming with the knowledge that comes from decades of experience in the field, and he was able to identify most of them at a glance, while I was still trying to get to grips with distinguishing between the more common North American bumblebees. There were black-tailed bumblebees,

[28] Harry H. Laidlaw Jr. has the limited but important claim to fame that he was the first person to work out how to artificially inseminate honeybee queens – a rather fiddly business as you might guess, but one which is enormously useful if you wish to selectively breed honeybees for particular traits.

Vosnesensky bumblebees, Van Dyke's bumblebees, massive iridescent purple-black carpenter bees and numerous smaller mining and leafcutting bees – over 1,000 species of bee have been recorded in California, and it seemed to me that most of them were living here. As if the bees weren't enough, hummingbirds also flashed through the shrubs, piping shrilly. Sadly, of course, there were no Franklin's bumblebees. Robbin proudly showed me a black-tailed bumblebee nest in what looked like a tit box nailed to a tree. This species is closely related to our early bumblebee, and is just as docile, for the bees were quite unperturbed when we lifted the lid and had a quick peek inside.

Once I had done my duty in Davis it was time to go on an adventure. Northern California in the spring is an absolute joy for a naturalist; the winter rains mean that there are flowers and lush greenery everywhere, while from April onwards one can almost guarantee sun every day until October, except in the highest mountains. The region has extraordinary geological and climatic diversity. The mountains on the west coast are bathed in mists and fogs caused by the contrast between the icy ocean and warm air, and they support spectacular sequoia forests which I had been lucky enough to visit before. Further inland and heading east, these forests give way to Mediterranean scrub, then to the agricultural plains of the Central Valley, and finally to the mighty Rocky Mountains.

Although the old haunts of Franklin's bumblebee are due north from Davis, I decided first to head north-west to the coastal mountain range, where the University of Davis has its own sizeable nature reserve. The McLaughlin Reserve comprises 7,000 acres of native Californian grassland and chaparral, habitats that have been all but entirely lost elsewhere. Paul, the reserve warden, was kind enough to take me on a tour. McLaughlin's used to be a gold mine, but when the gold ran out the land was handed over to the University of California, perhaps because the area was of little use for anything else. Parts of the site are badly contaminated from the mining operation. Gold mining is one of the most environmentally damaging of industrial activities, and for every 0.3 ounces of gold – enough for one ring – about twenty tonnes of toxic waste contaminated with mercury and cyanide is produced. By all accounts the company that mined at McLaughlin were better than most at minimising the environmental damage, and most of the reserve area was not badly damaged by the mining. However, McLaughlin is unusual for it lies on serpentine rocks which are naturally rich in toxic metals. As a result, the soils contain high concentrations of magnesium and iron, which most plants cannot tolerate. Most native Californian grasslands have been lost to agriculture, but those that were not converted to farmland have suffered heavily from invasion by European grasses which seem to have taken well to the Californian climate and outcompete almost all else. The serpentine soils at McLaughlin offer some protection against these invaders. Over the millennia, a lovely selection of native flowers have adapted to thrive on these soils, and these communities have been less invaded by European weeds because the invaders struggle to cope with the high levels of metals in the soils. Nonetheless some do, and the invading species are themselves slowly adapting to the local conditions, so Paul fights an ongoing battle with these aggressive weeds.

With 7,000 acres to look after, hand weeding is clearly not an option, and so Paul reluctantly relies upon graminicides – herbicides that selectively kill grasses – to battle the incoming grasses. This is not ideal on a nature reserve, but it is hard to conceive of an alternative that would work on the scale needed with the very limited manpower available to him. Fortunately the native grass species seem to have a little resistance to these herbicides, and although they are knocked back by them they do not die out entirely. However one feels about the ethics of using pesticides on nature reserves, it seemed to be working. Paul showed me untreated areas with few flowers and dominated by rank, dry grasses, while nearby the restored, pesticide-treated grasslands were spectacular, filled with no less than four different species of lupins in blue, yellow and cream, plus the dramatic blue spires of wild delphiniums, all emerging from a carpet of delicate native yellow clovers.

We spent a pleasant day, checking on the status of the invading weeds and the restored areas, and crawling around in narrow creeks searching for red-legged frogs, a nationally endangered species for which there were unconfirmed sightings at McLaughlin which Paul was keen to verify. Of course I obsessively caught and identified any bumblebees that I saw with an optimistic eye to turning up a previously undiscovered population of Franklin's bumblebee. We both failed – no Franklin's or red-legged frogs – but I did find a nest of Californian bumblebees, the workers pouring from a crack in the dried earth which must have led down to an old rodent burrow. This species had not previously been recorded on the reserve, so I felt that this indulgent day pottering about looking for flowers, frogs and bees hadn't been entirely wasted.

From McLaughlin I headed north – it was time to visit the old haunts of Franklin's bumblebee. My route took me up the middle of the Central Valley along Interstate 5, through endless flat fields of crops in neat rows, and past thousands of acres of almonds. This isn't the biggest almond-growing region in California – that is a little further south, closer to San Francisco – but it was still almond growing on a staggering scale. Farming in Britain, even in the arable heartland of the Fens, seems tinpot compared to this sort of industrial food production. Let me give you a few numbers – there are 800,000 acres of almond trees in California, producing 80 per cent of all the almonds grown in the world, about five billion dollars' worth, or 700 billion individual almonds. That is a lot of nuts in anyone's book.

As in the agricultural belt west of Buenos Aires, it is questionable whether producing food in this way is desirable or sustainable in the long term. In California, their intensive farming system is putting huge strain on the environment, and the cracks are beginning to show, quite literally. Each individual nut requires about five litres of water to produce, so that almond farmers are currently using 3.5 billion cubic metres of water on their crops. Of course, most other crops also require lots of irrigation water – I drove past huge fields that had been flooded, creating vast, square, shallow lakes, so that the soil would be moist for sowing melons, corn, tomatoes, peppers and potatoes. We humans also use vast amounts of water in our homes, gardens and on our golf courses, and in California it hasn't rained much in recent years, so that there just isn't enough to go around. Some of the almond farmers have rights to a share of the dwindling supply of river water for irrigation, but others do not and have resorted to drilling boreholes deep into underground aquifers and pumping up the water. Well-drillers are rushed off their feet, and there is now a waiting list of over a year for their services. When before they had to drill down perhaps 150 metres to hit water, now they often have to drill twice as deep, which inevitably makes both the drilling and the pumping up of the water more expensive. It is also, of course, a worrying sign – the aquifer water, which may have sat there for thousands or even millions of years, is being depleted. There is no regulation or monitoring of how much water is pumped, or how many wells are dug, so this new Californian gold rush is living on borrowed time. This isn't just a problem in California – across the USA, a jaw-dropping twenty-five cubic kilometres of water are pumped from the ground each year.

Aside from the obvious issue that aquifer waters will eventually run out, there are other problems with their use. This underground water contains lots of dissolved salts, absorbed from the surrounding rocks over the millennia. California being a generally warm and sunny place, the water evaporates fast from the soil or from the leaves of the crop, but the salts have nowhere to go and so they are building up in the soil, stressing and eventually killing the crop. Many farmers are aware that their almond crops are showing the early signs of stress, which include sickly, pale leaves lacking chlorophyll, and browning of the leaf edges. There is no easy way to get rid of the salt – heavy rainfall will slowly wash it away, but there has been precious little of that in the last few years.

If all that isn't enough to give pause for thought, the depletion of the aquifers is actually causing the Central Valley to sink. According to NASA in 2015, the land is falling by five centimetres a month in some places, and in total has dropped by up to two metres, causing damage to houses, roads, irrigation channels and bridges.

The vast orchards have other demands besides water. As I've already mentioned, the almond farmers ship in honeybees in February and March to pollinate their crops. It requires about 85 per cent of all the commercial honeybee hives in the USA to pollinate them – 1.7 million hives, or about eighty billion individual bees, which are brought in from all over the USA to pollinate the 2.5 trillion almond flowers. For most of these bees, this is but one stop on a long annual migration. After the almonds, they may be taken north to the apple and cherry orchards in Washington State, then east as summer comes to the fields of alfalfa and sunflowers in North and South Dakota, onwards to the pumpkin fields of Pennsylvania in August, and then finally to Florida for the winter. Others follow different routes, taking in squashes in Texas, cranberries in Wisconsin and blueberries in Michigan or Maine. Some of these hives cover 20,000 kilometres per year, but the almond orchards are by far the most lucrative stop for the beekeepers. This is the biggest commercial pollination event on the planet, and beekeepers can charge up to $200 per hive for the two to three weeks of pollination they provide. Just as the levels of the aquifers have gone down, the price of renting honeybee hives has gone up, for they cost just $75 ten years ago. The rise in price has been driven by two factors: an increase in the area of almonds, which has driven up demand, and the problems that commercial beekeepers face in keeping their bees alive, which has pushed down supply. The poor honeybees are stressed from all sides. They have to deal with the dozens of pesticides applied to all of these different crops, cope with outbreaks of foreign diseases and parasites, and put up with a bizarre, monotonous diet consisting entirely of almonds one month, cherries the next, and sunflowers after that. On top of all that they repeatedly spend days on end sealed in their hives and rumbling along on the back of a lorry, which must surely be confusing and stressful. Small wonder that beekeepers in the USA have been struggling to keep enough healthy hives alive to service the demand for almond pollination – Davis scientists have estimated that the average number of bees per hive arriving in the almond orchards has dropped from about 19,000 a few years ago to 12,000 today. Beekeepers are splitting their hives more often to make up for colonies dying, but the end result is that the colonies are getting smaller and weaker.

It is pretty clear that this whole system is teetering on the edge of collapse. If the drought continues, or honeybee problems get any worse, then Californian agriculture is going to be in big trouble, particularly the almond growers. Of course, one of these problems could be tackled by encouraging wild bees into the almond orchards. Neal's team have focused mainly on field crops, but incorporating areas of wildflowers or natural vegetation amongst the almond orchards to provide food and nest sites for native bees seems like a wise idea.

Although almonds are famous for being pollinated by honeybees, there is evidence that the honeybees' efforts are more effective when wild bees are also present. Claire Brittain, a postdoc working with Neal Williams and Claire Kremen, recently showed that honeybees actually change their behaviour when other bee species are present in the orchards. In almond orchards far from native vegetation, where no other bees occur, the honeybees tend to move very short distances between flowers. In orchards nearer the edge of the Central Valley, where mining bees and other native species often crop up, on average the honeybees move about more, and more frequently fly from row to row amongst the trees. Perhaps the honeybees are trying to avoid competition, so when they smell the odour of a native bee they skip on a bit (I'm guessing on this). You might wonder why it matters how far the honeybees fly – the answer lies in the planting arrangement of the orchards. Farmers tend to alternate rows of different almond varieties, and just as with apples, the different varieties cannot pollinate themselves; they need pollen from a different variety to produce a nut. So only when bees move from row to row do they effectively pollinate. Claire estimated that the wild bees increase yield by 5 per cent or more – not a huge amount, but with almond sales worth five billion dollars, a 5 per cent increase is not to be sniffed at. Of course, if something were to happen to the honeybees, those native bees would suddenly become worth a whole lot more. If I were an almond grower, I'd be making a bit of space for them.

The landscape was pretty desolate and I hacked northwards, stopping only for a quick coffee in a roadside diner which had the unusual feature of a stuffed grizzly bear that reared up menacingly in the corner, until the Central Valley narrowed and the mountains began to close in from both sides. From here the road began to climb, and Interstate 5 becomes rebranded as the 'Cascade Wonderland Highway'. It was certainly pretty, with endless rolling conifer forests on both sides, and the snow-capped peak of Mount Shasta, which I had seen from the air just a few days before, looming up in the distance ahead. For much of the approach to Mount Shasta the road follows the Sacramento River, the water rushing southwards in a hurry to irrigate all those almonds in the Central Valley. Just south of the great volcano the road veers westwards, side-stepping the mountain, and passing through the remote and unusually named town of Weed, where I stopped for lunch. The place was memorable only for the 'Weed like to welcome you' sign on entry to the town, but nonetheless I was sad to hear that much of it burned down in a wildfire in 2014. After Weed the road wound on through the forests for another hundred kilometres until I hit the Oregon border, and from there it was only another fifteen kilometres or so downhill to Ashland.

The ancestral home of Franklin's bumblebee turned out to be an old-fashioned country town with an historic heart of clapboard wooden houses, nestled amongst flower-filled meadows and set off beautifully by the snow-capped mountains to the east. Unfortunately for me a Shakespeare play was being performed in the local open-air theatre, and as a result the hotels in town were full. Ashland has an appealingly arty, hippie feeling, as if someone has picked up a piece of San Francisco and dropped it in rural Oregon. I ended up lodging in an edge-of-town motel, but its lack of charm was more than compensated by the lovely views of the surrounding mountains. I spent the next few days hiking the local area, net in hand, in optimistic search of Franklin's bumblebees.

There were a few more hazards than I was used to contending with on bee hunts back home. Dashing about through the undergrowth after bees in this area is unwise, as poison oak turned out to be common. Luckily Paul at McLaughlin had pointed out what it looked like, and had warned me that contact with the shiny green oak-like lobed leaves of this shrub can cause the skin to blister into sores that take weeks to heal, so I did my best to avoid it and only ended up with a few minor stings. More intimidatingly to someone from Britain, rattlesnakes turned out to be common along the rocky trails. The first one I came across I heard before I saw – a dry zither, which I foolishly mistook for the call of a cicada, and headed through the scrub to investigate. It was probably a good job that I was moving slowly to avoid any poison oak and so I spotted the fat brown snake before I could blunder too close. It eyed me with the fixed, defiant glare that all rattlesnakes have, and rattled its tail at me in warning. At this time of year they are just waking up from hibernation, and so they tend to be sluggish – most I subsequently saw were just poking their triangular heads from their burrows, and the first I noticed of them was the movement when they darted back inside.

It was a lovely area to explore as spring burst into life, and I felt like a child again as I wandered about, net in one hand and camera in the other, soaking up the sights and sounds. Bobwhite quails were common in early morning – charming, fast-running little creatures with an outlandish floppy crest, resembling someone from a New Romantic pop band of the Eighties. They would explode out of the bushes as I approached and career along the trail ahead like clockwork toys. Jackrabbits lolloped about more sedately, seemingly weighed down by their comedic, oversized ears. The trees were alive with skirmishing animals, perhaps establishing their summer territories, or perhaps just driven by the surging hormones of spring, so that the forest rang with a cacophony of squawks. The western scrub jays and grey squirrels seemed to have declared war on one another and were engaged in running battles in the canopy, presumably contesting shared food sources, though from what I could see neither side had the weaponry to achieve a convincing victory. Meanwhile, rival gangs of acorn woodpeckers pecked hell out of each other – these are unusual birds in that they live in family groups, sharing the work in looking after a single nest, and storing thousands of acorns for the winter in a 'granary', a tree that they drill with a myriad individual acorn-sized holes. The acorns dry and shrink over time, and the birds occupy the long winter months by constantly shifting acorns into more appropriate, tight-fitting holes. Presumably these battles were over the last of their winter stores – on one occasion a pair that were locked in combat fell from a tree and crashed to the forest floor not two paces in front of me. Their feet were entwined, and each seemed intent on pecking the other's eyes out, so I was relieved when they finally noticed me and shot off in different directions, apparently unharmed.

Bumblebees were plentiful. There were most of the species I had found with Robbin, plus more that I had not seen before, including the delightfully named fuzzy-horned bumblebee, a shaggy-haired little bee with no discernible horns. Of course I didn't find any Franklin's bumblebees, despite catching and inspecting hundreds of different bumblebees, many of them queens fresh from hibernation. After failing to find any *Bombus dahlbomii* in Argentina, I should not have been surprised, for this was always going to be the longest of long shots, a ghost hunt for a bee that local experts had failed to find for six years. Once I got myself briefly excited, but quickly realised that I had captured a lovely queen of the rather similar 'obscure bumblebee'.

After a few days it was time to pack my walking boots and head back south to fly home. I took the scenic route, skirting east into the Cascades and then southwards to the Sierra Nevadas, passing through some truly wonderful scenery though I hardly had time to appreciate it. On my last evening in California I stayed in a shabby little motel on the southern shores of Lake Tahoe, an idyllic, emerald lake perched 2,000 metres up in the sierras on the

Nevada border, due east from Sacramento. California's rich come here to party; the shoreline is a little spoiled by casinos, in one of which Frank Sinatra occasionally used to sing. JFK and Marilyn Monroe are said to have had an affair in a cabin in the woods, though I'd hazard a guess that it was a rather nicer cabin than the one I was staying in. I sat on a rustic bench among the tall firs, and feasted on rye bread, tomatoes and a deliciously tangy local goat's cheese named Humboldt Fog, though I had my work cut out to repel the bold advances of a gang of chipmunks who had designs on my comestibles. Terrapins slowly hauled themselves out onto rocks along the lake shore to catch the last of the day's sun. Beyond the lake to the north and east, I could see endless serried rows of snow-capped peaks and forested valleys, fading into the distant haze. The Rocky Mountains cover a vast area, much of it inaccessible, and there are not too many entomologists able to recognise a Franklin's bumblebee, should they be lucky enough to stumble across one. Perhaps Franklin's bumblebee now exists only in Robbin's memory, and as a few pinned specimens in his office at Davis.[29] Or perhaps not. There could be some left, somewhere out there, in a valley that the disease never reached, or maybe the offspring of a few individuals that just happened to have a little natural resistance. I'm sure that Robbin will continue to diligently search its old haunts until he himself loses his personal battle with extinction. It would be wonderful if he could see just one more.

[29] In 2010, Robbin and the Xerces Society for Invertebrate Conservation submitted a petition to the US Fish and Wildlife Service to list Franklin's bumblebee as an endangered species. The Fish and Wildlife Service are so snowed under with litigation and petitions that, five years later, they have not yet had time to assess the case, so Franklin's is not yet formally recognised as being endangered, let alone extinct.

CHAPTER SIX

Ecuador and the Battling Bumblebees

May your trails be crooked, winding, lonesome, dangerous, leading to the most amazing view. May your mountains rise into and above the clouds.

Edward Abbey

Why do students move so slowly? This is a question that has vexed me for many years. Every university building I have ever worked in has possessed long corridors, which in termtimes are choked by gaggles of students ambling along purposelessly. True, some are more than a touch overweight, and this must slow them down, while others are 'Facetiming' friends or chatting on their hands-free iPhones, which presumably distracts them from the complex business of placing one foot in front of the other. But even those who aren't plugged in to the internet or suffering from a surfeit of KFCs often seem to move at an almost imperceptible pace. Perhaps they just have too much time on their hands and we academics are to blame for not giving them more work to do. No matter the cause, I find their trundling irksome as I am generally in a hurry and they provide an endless moving chicane of obstacles to negotiate. This may sound odd, but I often run from place to place, even along the corridors at the university, and almost always when going across campus. I particularly like to run up stairs whenever the opportunity arises, and I find it enormously frustrating when I come up behind a phalanx of large-bottomed students blocking the stairwell. Life is too short for such dawdling.

Given my impatient and perhaps entirely unreasonable attitude to ambulation, it was particularly humiliating to my foolish macho pride to find myself at the rear, struggling to keep up with a column of students as we wound our way up a steep muddy path high in the Andes in September 2014. The bus had dropped us off at the closest point it could reach, about six kilometres of jungle trail from our destination, and so the thirteen students, two other staff and I were slogging up the slippery path. A team of four mules was carrying our rucksacks and equipment, which had been strapped in improbably large bundles to the sturdy beasts' backs. Perhaps it was age, or the altitude, or jet lag, but whatever excuse I could think up did little to salvage my injured pride as the last of the students disappeared into the misty forests above me, leaving me panting and sweating far behind as the trail climbed up into the clouds. It turned out that I was just about to go down with food poisoning, the likely result of consuming a large, meat-filled pastry from a roadside stall in Quito the day before; it left me weak as a kitten for the next few days.

We were here for a two-week residential field course from the University of Sussex, in which the students were to learn about the wonderful biodiversity and ecology of the cloud forest, high-altitude rainforests perched between 1,500 and 4,000 metres up on the precipitous sides of the Andean mountains. These forests are famous for their huge diversity of birds, orchids, butterflies and much more. Ecuador is one of the world's great biodiversity hotspots, bisected by the spine of the Andes which run roughly north to south. To the east of the mountains, Ecuador encompasses a portion of the steamy Amazon basin, a part as yet affected relatively little by deforestation. The west is more densely populated, with fragments of tropical forests amongst farmland, and palm-fringed beaches that attract tourists looking for an alternative destination to the Caribbean. Far off the coast to the west, the Galapagos Islands are also part of Ecuadorian territory. The Andes themselves support a huge diversity of habitats, from bleak ash-strewn active volcanoes capped with snow to lush green highland cloud forests and chilly paramo grasslands above the treeline. Although not much larger than Great Britain, Ecuador supports about 10 per cent of all the known species of animal and plant on Earth. The figures are astonishing – 317 mammal species, 460 amphibians and 410 species of reptile (the respective numbers for the UK are 101, seven and six). Of course no one has a clue how many insect species there may be in Ecuador, but the butterfly list so far runs to about 4,500 (compared to about seventy in the UK).

I sat down for a rest at the side of the track, gasping for breath with sweat trickling down my spine. Clearwing butterflies flitted ghost-like in the shade beneath the forest canopy, their wings largely devoid of coloured scales so that they are transparent aside from a delicate network of dark scales and veins. A long-winged *Heliconius* butterfly,[30] black with golden-yellow stripes, soared effortlessly along the path and fluttered around me, perhaps briefly wondering if my red T-shirt was some new exotic flower. I noticed a large web-lined burrow in the bank at the side of the path, from

---

[30] These elegant creatures have a trick that enables them to be amongst the longest-lived of butterflies, with the adults being on the wing for three months or more. Most butterflies can only drink nectar, but *Heliconius* butterflies also gather pollen on their proboscis, exuding enzymes to digest it and then sucking up the resulting protein-rich soup. In some *Heliconius* butterflies the males seek out female pupae and mate with them before they have hatched into adults, which seems like a morally questionable tactic.

which the front two hairy legs of a sizeable tarantula protruded. Big spiders have always given me the willies, so I decided it was time to move on.

I eventually climbed above the cloud and emerged at our destination, a wooden lodge perched in a clearing on the top of a mountain at about 1,800 metres in altitude. The views were breathtaking: densely forested peaks rose from a woolly blanket of white cloud that shrouded the valleys far below. Hummingbirds whirred amongst the pretty shrubs around the lodge, exotic birds called from the surrounding forest, and colourful butterflies soared on the thermals.

Santa Lucia Cloud Forest Reserve is a community endeavour, created by a dozen or so local families who were struggling to make a living as small-scale farmers, and so decided on a different approach. They pooled their land to create a large nature reserve which now protects 735 hectares of forest. To make a living, they constructed the lodge for visitors, somehow lugging up everything they needed on the mules or on their own backs; how they got the glass for the windows up there in one piece I couldn't imagine. They struggled for a while to make a go of it, with few visitors, until Mika Peck came along. Mika is a conservationist at Sussex University, a boyish forty-something with an infectious sense of humour who spends his life championing various rainforest conservation projects in Ecuador and Papua New Guinea. He got funding from an organisation called Earthwatch to bring a group of volunteers out to Santa Lucia to study the wildlife. The volunteers had to pay quite a bit for the privilege, and a chunk of this got passed on to the Santa Lucia folk, quite a windfall for their embryonic business. Mika and an Earthwatch team came back annually for four years, and the money was ploughed in to improving the lodge, building some more comfortable accommodation in satellite huts, and creating a working hot-water system (no small feat on the top of a mountain in the middle of nowhere). Mika then started running an undergraduate field course for students from Sussex University at Santa Lucia, and a couple of other universities followed suit, bringing in another reliable income stream. The future is far from certain, but for the moment they seem to be keeping afloat.

One of the fascinating things about the place is that it is run as a cooperative, with no one in charge. The numerous joint owners and their families all chip in, doing repairs to the plumbing, leading guided walks, cleaning the toilets – whatever needs doing, and seemingly always with a cheerful smile. Mika did confess that behind the scenes it can be pretty chaotic, particularly when it comes to making big decisions, but on the whole it seems to run remarkably well.

The forests of Santa Lucia are home to an impressive range of endangered mammals, including a good population of spectacled bears, the species to which Paddington presumably belongs (although he hails from neighbouring Peru). The bears have suffered greatly in recent decades, from loss of their habitat and illegal hunting – grotesquely, their paws sell for $20, while their gall bladders are highly valued in Chinese medicine (along with a seemingly random selection of body parts of other highly endangered animals), and can fetch $150 each. Given that the average monthly wage in Ecuador is just $30, it is easy to understand why an unscrupulous person with a gun might be tempted to shoot one.

Other mammals to be found in the forests of Santa Lucia include the jaguarundi, ocelot, oncilla, margay, tayra, kinkajou and the potentially dangerous puma, the largest remaining big cat (sadly, jaguars were hunted out in this region years ago, but the owners of Santa Lucia dream that they might one day return).

As you might imagine, the students were particularly excited at the prospect of seeing such amazing and exotic wildlife, perhaps even sighting a puma or a bear. Most had never had the opportunity to visit the tropics before. I, however, had an ulterior motive for being there. One of the other staff, Jeremy Field, who had visited this site many times before, had mentioned that there were bumblebees to be found. Jeremy isn't a particular expert on bumblebees – he studies 'primitively social' wasps and bees, those that teeter on the edge between solitary and social life[31] – but he knows a bumblebee when he sees one. Almost nothing is known of the ecology or behaviour of the bumblebees that live in tropical South America; most species are known only from a few pinned specimens in museums. Jeremy had also mentioned that there were orchid bees to be seen, a group of spectacular, large and colourful bees found only in the neotropics. So a chance to see and study such little-known creatures, whilst also doing my day job – teaching students – was far too good to resist.

When we arrived at the lodge in the late afternoon we were served the most delicious cup of thick hot chocolate, made from cocoa beans grown in the valley below. It turned out to be a wonderful place to stay, both primitive and yet comfortable. Soon after we arrived it began to get dark, and in the tropics the transition from day to night is very fast as the sun seems to drop like a stone from the sky. We sat on the balcony as darkness fell,

[31] All bees, ants and wasps share a common ancestor that lived about 240 million years ago and was a solitary creature. Since then, full sociality in which one or more queens are aided by sterile workers has evolved at least eleven separate times, giving us honeybees, bumblebees, ants, common wasps and so on. How and why this has happened so many times in this one group of related insects remains something of a puzzle, and so studying those bees and wasps which are halfway between the two states might provide some clues as to the answers.

listening to the nocturnal creatures of the forest as they woke up and started serenading one another, or staking a claim to their territories. We could only speculate as to what made the many and various calls, some beautiful, eerie and haunting, others just insistent, incessant buzzes and rattles. Were they crickets, cicadas, frogs, nocturnal birds such as nightjars or potoos, or some other exotic creatures unknown to us? Fireflies began to flicker amongst the trees, and bats launched themselves from roosts under the eaves of the building. It had been more than a decade since I'd previously visited the tropics, and it was magical to be back.

There was no electricity, so we ate dinner by candlelight. The food was simple, rice and beans, but delicious nonetheless, particularly after the long, exhausting hike up through the forest. I could easily have eaten twice as much, and a cheese course wouldn't have gone amiss, as the stomach bug hadn't yet struck. It turned out that most of the ingredients for our food were grown in an organic garden next to the lodge; otherwise they were bought in the market of the nearest village, some eight kilometres distant, and had to be carried up by mule. This was proper, seasonal, low-food-miles eating – the nearest supermarket might as well have been on the moon. We went to bed early, exhausted, and excited about what we might find the following day.

I was woken the next morning by the purr of a hummingbird feeding on the flowering shrubs outside my window. The clouds had cleared from the valleys, and the sun was peeking over the mountains to the east. After a quick breakfast of fruit and yogurt washed down with lots of coffee, all of which I was to become reacquainted with later that morning, we set out on a walk in the forest to get our bearings. Mika knows these forests like the back of his hand, and he took the lead along a narrow, winding forest trail – I followed at the rear so that I could stop to look at anything that caught my attention. Huge, buttressed trees towered above us, festooned with epiphytes – plants that grow on other plants rather than having their roots in the soil. The branches of the trees were laden with orchids, bromeliads and ferns in enormous diversity; there are no less than 2,500 species of orchid alone in Ecuador. The moisture brought in by regular immersion of the forest in cloud allows these epiphytes to suck up enough water to thrive. The bromeliads have an additional trick – their leaves form gulleys that catch rainwater and feed it into a central well, so that each plant has its own tiny private water reservoir. These mini aerial ponds are themselves home to a multitude of wildlife, from hoverfly larvae to tadpoles.

In no time at all I spotted my first bumblebee, a male. I was alerted by a familiar buzz – lower in pitch than most other insects in the cloud forest. My movement had disturbed a bee perching close to the path, so I froze to see what it would do. It soon settled again, on the tip of a heart-shaped leaf of some low-growing forest plant, a foot or so above the ground, in an area of dappled sunlight at the side of the path. Most of the ones I found subsequently were in similar places. It didn't look much different to familiar British species – about the same size and shape, black with two yellow stripes and a white bottom. If one were to chop off the front half of a garden bumblebee and glue it on to the back half of a tree bumblebee you'd have a pretty good imitation (though I guess that won't mean a lot to most folk). I didn't know it at the time but subsequent investigations – otherwise known as asking the bumblebee taxonomy guru, Paul Williams, at London's Natural History Museum – revealed it to be a species named *Bombus hortulanus.*

The behaviour of this bee was very different to anything I had seen before. The bee was crouched, his antennae cocked forwards, his abdomen pulsing, in an alert, aggressive posture most unlike the generally relaxed behaviour one associates with bumblebees. He was restless, shifting his position on the leaf feverishly every few seconds, and dashing out at anything that flew past. I soon realised that he was not alone – two other males were perched nearby, only a metre or so apart. They would chase any other flying insects that came near, and if they strayed too near to one of the other males in their foray they would be attacked, the two bees then swirling about in frantic aerial combat for a few seconds before each returned to their perches. They did not always sit in exactly the same place – each seemed to have a few favourite leaves, all close together, and they would alternate between them.

I realised that the rest of our party had left me behind, so I quickly marked the spot with some coloured ribbon tied to a twig, and ran to catch up, while beginning to feel queasy. I pondered what I had seen – it was presumably some sort of behaviour related to finding a mate. The mating habits of bumblebees are somewhat enigmatic. Darwin was one of the first to study them, in his garden at Down House in Kent. The males of many species, including the garden bumblebee studied by Darwin, mark out circuits perhaps 200 metres long with pheromones, and then streams of them zoom round and round the circuit in the same direction like little sports cars with their engines revving, presumably hoping to impress a female. Oddly, virgin queens rarely if ever show any interest in their boy-racer antics. In other species, such as the tree bumblebee, the males are more direct; they hang around in excitable crowds outside nests that are producing new queens and simply try to grab them as they emerge. Males of a few bumblebee species gather on the tops of hills and there await the arrival of virgin queens, though once again no one has actually observed a virgin arriving. But I had seen all of these behaviours many times before, and the Ecuadorian males were clearly doing something altogether different.

I recalled that there is a fourth behaviour, described from a handful of North American and Asian bumblebee species such as the Nevada bumblebee (*Bombus nevadensis*), in which the males are fiercely territorial. In these species, the males are said to have distended, oversized eyes, helping them to spot any queens that hove into view. These boggle-eyed beasts fight over prominent perches, and use them as lookout posts from which to scan for potential mates and keep a beady eye on rivals. If they spot an incoming virgin queen they dash out to intercept her mid-air like scrambling fighter pilots, drag her to the ground and unceremoniously attempt forced copulation. Were these Ecuadorian bees adopting a similarly charmless strategy? They didn't have boggle eyes, and didn't seem to be perching in prominent positions, but otherwise the behaviour looked similar.

Over the next few days Jeremy and I were busy teaching the students how to identify insects and I was rushing to the bathroom every five minutes (composting toilets may be great for the environment but in tropical heat they aren't somewhere you want to go often when already nauseous) so I had little time to devote to investigating the bees' behaviour any further. With the students we forayed through the forests, armed with butterfly nets and sweep nets (the former for catching flying insects, the latter designed for bashing through undergrowth to catch resting insects). Just as with the kids at the Dunblane primary school, it was fun teaching the students how to use these nets – it is all too easy to flap around excitedly and achieve nothing other than frightening all the insects away. We explained to the students how to distinguish the different insect types, such as grasshopper, beetles, wasps or mantises. These are crude divisions that each encompass huge numbers of species, but identifying the insects any further was usually impossible. In the UK, we are spoiled with guides and keys that enable us to identify any insect to species. Our fauna has been described and studied in detail, and very few new species are likely to be discovered. In the tropics, however, any sweep of the net is likely to catch new species that have never been formally described by scientists. It is estimated that we have only named perhaps one-fifth of the species on Earth, and many of the four-fifths remaining are likely to be insects that live in tropical forests. Catching new ones is easy, but identifying which of the insects in a net are the new ones is enormously difficult. Sadly, there are a small and dwindling number of specialist insect taxonomists with the vast knowledge needed to distinguish anything new from the ones we already know about. Nobody seems to want to fund this sort of work any more.

Regardless of the ongoing effects of the Quito pastry, it was wonderfully exciting dashing about looking for insects in the forest. Jeremy was particularly funny; at university he is a quiet, gently spoken and unassuming chap, a well-respected professor and expert in the evolution of social behaviour in insects. In the field he was transformed with excitement, sprinting about with his butterfly net with all the enthusiasm of a ten-year-old – clearly he never grew out of his bug period. He has the added advantage of being enormously tall and gangly, able to snaffle high-up insects that I could not possibly reach, so I had to pull out all the stops to match him as we vied with one another to catch the most interesting specimens. We leaped and chased, turned over logs and stones, sifted through dung, and waded about in streams in search of interesting beasts. Between us we found all sorts of curious and marvellous creatures; I caught a dobsonfly, a rare, giant, primitive-looking relative of the lacewing with enormous but actually rather feeble jaws. Jeremy countered that with a bess beetle, a large, shiny, black creature that squeaked – apparently they communicate with their grubs in this way. I trumped that with a magnificent owl butterfly the size of a small bird, with huge eye-spots creating a life-size replica of the face of an owl. Jeremy went one better with a terrifyingly large tarantula hawk wasp, a velvet-black creature the size of my thumb that feeds its offspring on paralysed tarantulas (we were later lucky enough to see one dragging its helpless, limp prey – a rather beautiful blueish tarantula – back to its burrow).[32] And so we went on, catching a fantastic array of insects, including slender stick insects, spiny, camouflaged bush crickets, moths that beautifully mimicked yellowing leaves, scurrying cockroaches, caterpillars adorned with outlandish spines and protuberances, and much more besides.

For me, perhaps the most exciting insects we saw, aside of course from the bumblebees, were the orchid bees. There were two

[32] Heroically or idiotically, depending on your point of view, an American entomologist named Justin O. Schmidt deliberately got himself stung by seventy-eight species of insect so that he could rank and describe the pain they produced. The tarantula hawk wasp came out joint top with the bullet ant. He colourfully described the pain as 'immediate, excruciating pain that simply shuts down one's ability to do anything, except, perhaps, scream. Mental discipline simply does not work in these situations.' His description of the sting of the bullet ant, a giant ant species also native to South America, was 'pure, intense, brilliant pain. Like fire-walking over flaming charcoal with a 3-inch rusty nail in your heel.' Clearly both are best avoided. Schmidt's work was published in 1990, and subsequently inspired his fellow entomologist Michael L. Smith to deliberately get stung by honeybees on twenty-five different parts of his anatomy, his goal to find out which parts were most sensitive. It turns out that the most painful parts of the body are the nostril, the upper lip and the penis. For their magnificently selfless if somewhat pointless endeavours, Schmidt and Smith were jointly awarded the Ig Nobel Prize in 2015, a parody of the Nobel Prize scheme.

types – one, a *Euglossa* species, shiny, iridescent green, a little smaller than a bumblebee. Like miniature hummingbirds, they hover by flowers without landing, using their long tongues to probe for nectar. They have a swift, darting flight, punctuated by pauses when they hover in mid-air, surveying their surroundings, their bodies absolutely motionless but their wings a blur. Interestingly, a local fly species has evolved to mimic both their metallic colour and their flight pattern, presumably to fool birds into thinking it also has a sting.

The second type of orchid bees were much larger and densely furry, with orange, black and yellow stripes – a species belonging to the genus *Eulaema,* according to Jeremy. When I first saw one I was fooled by the size and furry coat into thinking it was a bumblebee queen, but close up it was clear that this was no bumblebee – it had greatly enlarged thighs on each of its hind legs, something that all male orchid bees share. These oversized legs are hollow, with a small hole into the interior, and they are used to store fragrant, volatile chemicals: odours that are needed to attract and successfully woo a female. Males of most orchid bee species collect these chemicals from just one species of orchid, and the orchid has come to rely on the particular bee for pollination. In the perfume industry, a technique known as enfleurage is used to trap elusive scents from flowers by absorbing them into fat or grease. Male orchid bees use this same process; they excrete a drop of fatty liquid from the labial glands in their head onto the orchid flower, and once it has absorbed the scents of the flower they scoop it up and store it in their legs. The fats are then absorbed back into the body (without the floral scents) and recycled to the labial gland for use on the next flower, so gradually concentrating the floral scents in the leg chamber. Sometimes males have been seen to mug one another to steal these chemicals, pinning down a rival and sucking the fatty perfume from their legs rather than bothering to visit the many flowers needed for them to collect legs-full for themselves. Nests of orchid bees are exceedingly hard to find in these tropical forests, but their social life is said to be intermediate between solitary and social bees. They are thought to live as small groups of females, each of equal status, and all laying eggs – right up Jeremy's street, for they might give us a glimpse of an early stage in the evolution of much more complicated insect societies such as those of honeybees or ants – if only one could find the nests.

While I was getting all excited over orchid bees, Jeremy's most prized find was two nests of some minuscule wasps called *Microstigmus*. These little creatures are nothing much to look at, being just three or four millimetres long, but they evolved a social lifestyle entirely independently of other groups of social wasp or bee. They feed on springtails and thrips, and live in a little nest made of silk, no bigger than a walnut and hanging from silken threads in a sheltered spot on a buttress of a tree trunk or under a large leaf. Jeremy had previously studied a different *Microstigmus* species in the Atlantic rainforests of Brazil and found that the males are actively involved in defending their nests, something that is extremely unusual in bees and wasps (usually the males are lazy creatures with just one role in life – mating). Hence he was keen to find out more about the life of this Ecuadorian species.

Despite our best efforts in capturing insects, we could not compete with Mika in impressing the students. He'd been out setting camera traps to capture images of the forest's mammals, and in no time at all he had photographs of spectacled bears, ocelot and even a particularly large male puma, the latter only a few hundred metres from the lodge. In his excellent book *Feral*, George Monbiot argues that those of us living in countries that are now devoid of dangerous wild animals subconsciously miss the frisson of excitement they generate, and I'm inclined to believe that he might be on to something. It certainly made life more interesting knowing that such magnificent creatures were living all around us, particularly when we went out at night by torchlight onto the forest trails to see what we could find. Suddenly, our imagination turned every rustle of a mouse into the snuffling of a bear[33] or the sound of a puma crouching to pounce.

Each night, I hung up a white sheet in the edge of the forest near the lodge and shone an ultra-violet lamp onto it. To our eyes these lights give out an odd purple glow, but we are unable to see most of the light that they emit. To insects they are enormously bright, for insects' eyes are able to detect ultra-violet light that is invisible to us. Within seconds of turning on the trap, moths in bewildering variety started to arrive. Some were small and unremarkable, subtly marked in shades of brown, sometimes with scalloped wing margins to give them camouflage when sitting on bark or dead leaves. Some had prominent spots and blotches, in cream, yellow, orange or red, looking very conspicuous on the white sheet but presumably affording them camouflage when perched on forest leaves blotched with fungus or the squiggled marks of leaf-miners. Some were huge: giant silk moths with feathered antennae that they use to sniff out females from miles away; fast-flying hawk moths with powerful, streamlined bodies and long, sharp-pointed wings streaked in green, brown and yellow, which bashed into the other insects as they whirred around the

---

[33] Spectacled bears are actually the most docile of bears, and naturally eat a largely vegetarian diet of bromeliads, palm hearts and berries. There is only one recorded case of a spectacled bear killing a human – the human concerned was a hunter, and had just shot the bear, which was up a tree at the time. The dying bear fell on top of the hunter and squashed him. Just occasionally in life (and death), justice can be satisfyingly swift.

light in excitement or confusion. Other insects came too – noisy cicadas lured in from the trunks of the forest trees where they normally sit to sing; huge chestnut-brown scarab beetles, flying with all the elegance and grace of a particularly clumsy house brick, would come crashing in to the sheet with their wings clattering, and then fall to the ground where they would fold away their wings and sit, apparently exhausted from their journey. Click beetles, wasps, flies, true bugs, and more arrived, some settling down quietly on the sheet, many continuing to zoom around in chaotic circles. It was a fantastic demonstration of nature's diversity, all displayed on one white bed-sheet. The students loved it, gathering around the sheet in the midst of this storm of insect life, pointing and exclaiming with delight as different weird and wonderful insects arrived, and not minding that the moths landed in their hair and occasionally fell down inside their shirts.

In addition to the animals, the students were also taught how to identify the local plants by Anna, a lovely Ecuadorian lady with an encyclopaedic knowledge of the mountain flora. The plants of Ecuador were all utterly unfamiliar to me, mostly belonging to families that do not occur in Europe, so I tried to pick up some knowledge from her as we went along. Even where the plant families were familiar, the plants themselves looked utterly different – for example in Ecuador, members of the daisy family grow into trees thirty metres tall. It was hard to see any resemblance to the daisies and dandelions in my lawn at home. Anna explained that a great many of the local plants are adapted for pollination by hummingbirds, having deep, tubular, often red flowers that have co-evolved with the elongated bills of the birds. These spectacular birds were everywhere in the forest, the thrum of their wings and their chirping an almost constant companion, though they were often hard to see clearly amongst the dense foliage.

At dawn each day Mauricio, one of the many locals who jointly own the lodge, would lead a birdwatching walk through the misty forest. Dawn is the time of peak bird activity, when they have a brief hour or so of frenzied calling before melting back into the dense canopy. Even then they can be very hard to spot high up in the trees but Mauricio was able to mimic the call of many types with uncanny accuracy, and so he would call back to them. Many were sufficiently intrigued by this new rival or potential mate that they would come down from the treetops to investigate, giving us a great look at them. The evocative names of some of the many species we saw speak for themselves: flame-faced tanagers, golden-headed quetzals, red-billed parrots, masked trogons, toucan barbets, plate-billed mountain toucans, and so on. Among these stunningly beautiful creatures, my favourite was the quetzal. Twenty years or more ago I spent many hours trying to spot them in the forests of Belize without any success, though we heard them in the distance. Here, the quetzals were quite tame, and it was worth the long wait to see them. These are plump, pigeon-sized birds, clothed in emerald green but with a red belly, and in sunlight their green head flashes iridescent gold. The Aztecs and Maya viewed the quetzals as 'gods of the air' and as symbols of freedom, goodness and light – their rulers wore headdresses made from the long green tail feathers of the males, which can be up to half a metre long, and were valued more highly than gold. Since the birds were thought to be incarnations of the god Quetzalcoatl and hence were sacred, they were caught live, their tail feathers plucked, and then released (which seems like a pretty undignified treatment for a deity). At Santa Lucia, one male quetzal was particularly bold and would often visit us at the lodge, settling in a tree on the edge of the garden and peering down at us with his gleaming head cocked to one side, as a god might look down curiously at the antics of the mortals below.

When we got back from the early-morning walk we would pause by the hummingbird feeders hung outside the lodge. These were filled with sugar solution, and drew dozens of these tiny birds from the forest for an easy feast. The birds were accustomed to humans, and so we could stand just a few feet away while Mauricio rattled off their names. Hummingbirds are impossible not to love – incessantly active, tiny aerial jewels, glittering in plumage of greens and blues with flashes of red, purple or bronze on the crown of their head, throat or tail, depending on the species. I do not think I will ever fail to smile when I glimpse a hummingbird. The different species were mostly distinctive; the male violet-tailed sylph sported tail streamers longer than the rest of his body. The booted racket-tail had fluffy white leggings and a pair of long wire-thin tail streamers with oval flags at their tips. Some species had stubby beaks, others long, straight tapered beaks, while the tawny-bellied hermits were equipped with scimitar-curved bills. Occasionally a little woodstar would put in an appearance, one of the smallest bird species in the world, little bigger than a bumblebee and flapping its wings at a frantic seventy times per second. This was close-range birdwatching for idiots – contrasting sharply with the peering through binoculars at distant small brown jobs that I was familiar with from birdwatching in the UK.[34] With Mauricio's help we quickly learned to identify them all. In total we saw no

[34] I shouldn't be rude about British birds – many are beautiful and all are fascinating – but I became disillusioned by birdwatching as a boy because I could never be quite certain which small warbler, lark or finch I had glimpsed, and I didn't have an ear for distinguishing between their songs. If you are tempted to take up birdwatching, I strongly recommend finding an expert who can help you, as otherwise you may be in for a frustrating experience. Alternatively, take a trip to Ecuador.

less than seventeen species of hummingbird on these feeders during our stay, and they fed from dawn to dusk, so whenever I had a spare moment I would pull up a seat and watch them, grinning foolishly to myself.

On the fifth day, the students divided themselves into small groups and set about devising their own original research projects, with a little steering from the staff. I ended up supervising two projects: one a group of three girls who wanted to study the behaviour of the hummingbirds, and the other a pair of girls who were keen on comparing numbers of butterflies and moths in primary forest versus cleared areas or secondary, damaged forests. Once the students were organised and had started collecting their data, I was able to turn my attention back to the bumblebees. I searched all of the forest trails that radiated out from the lodge. It was gruelling walking for the terrain was rugged in the extreme; the trails were all steep, often very narrow paths cut along the side of vertiginous slopes. Slipping off these muddy trails would not have been a good idea. Some trails wound up towards a nearby peak at 2,500 metres altitude, while others plummeted down into gorges at about 1,300 metres, where mountain torrents poured over huge rounded boulders and colourful butterflies fed on salts from the damp sand at the water's edge. Fortunately by then I was getting over the food poisoning, and my legs had a little more zip in them than before. It was exciting being off on my own in these remote forests, hunting for bumblebees but knowing that I might come face to face with a bear, puma or eyelash viper around the next corner.[35]

---

[35] Although of course I never did. The odds of seeing any of these creatures are slim, and if I had been lucky enough to glimpse one it would have been far more interested in melting back into the forest than in attacking me. In reality, going for a walk in any city is far

I found quite a lot of bumblebees, all male *Bombus hortulanus*, and all doing more or less the same thing as before: sitting on a leaf, anxiously twitching and pulsing, little balls of energy. Many were in small groups of between two and five, but some seemed to be on their own, fiercely defending their chosen patch of the forest floor even though nobody else seemed to want it. There appeared to be nothing remotely special about the places they chose to sit – they weren't especially sunny or shady, and there were no flowers – so far as I could see there were a million other spots that looked just as good.

The bees were fairly common on the trails uphill, right up to 2,500 metres, which was as high as the mountains within walking distance got, but they petered out quickly when I went downhill; I found none below 1,600 metres. I marked the positions of every one with red ribbon on the stem of their favourite perch so that I could find them easily on the following days.

I went back to watch them whenever I could find time, which was often, as my two groups of students seemed to be getting on well. The hummingbird girls were studying, amongst other things, whether boldness differed between species, and had found that some types of hummingbirds were very bold indeed; they had brown hermits and violet-tailed sylphs actually perching on their hands to feed. My bumblebees were much more flighty, and I had to approach each group of them cautiously. Fortunately their positions didn't usually change from day to day, so I knew when to slow down and creep forwards to watch them. Sometimes a bee

---

more dangerous than exploring the rainforest, though our familiarity with cars and muggers breeds contempt for the more realistic dangers they pose. Should you be wondering, the eyelash viper is a venomous, small yellow arboreal viper which sports a splendid pair of Denis Healey-style eyebrows.

would be missing from his post when I arrived, but if I sat still and waited for a while he would usually return. Occasionally I saw the bees wiping what I guessed to be pheromones onto their perches. Male bees have hairy moustaches, and some species, including those that Darwin observed in his garden all those years ago, use them like paintbrushes to mark leaves and twigs along their race tracks, using a pungent liquid exuded from glands in their head. Presumably the bees were marking their patches, either to attract females or to repel males.

If it rained or became heavily overcast then all the bees vanished – presumably heading into shelter somewhere – but as soon as the sun came out they returned. Occasionally one would disappear for half an hour, perhaps to feed on nectar, though I rarely saw them on flowers – my best guess is that they fed mostly on the flowers of some rainforest tree, far above my head. Presumably this is also the case for the *hortulanus* worker bees; I didn't see a single one, but given the number of males there must have been plenty of workers somewhere nearby. One of the great challenges for a biologist working in tropical forests such as this is that much of the action happens in the inaccessible canopy.

On just one occasion I saw a male *hortulanus* feeding on the flowers of a shrub belonging to the Rubiaceae. This is another plant family whose members are unrecognisable in South America; in Europe, the best known members are low-growing, scrambling plants of meadows and hedgerows such as lady's bedstraw and cleavers (obscurely known to generations of kids as sticky willy). In South America, once again many of the members of this family are large shrubs or trees, including coffee – making me wonder whether these Andean bumblebees contribute to coffee pollination in Ecuador, though I did not get a chance to check this out.

Where there were groups of males, their aerial battles were fascinating to watch. Sometimes they went on for minutes – two bees often flying headlong at one another like teenage joyriders playing chicken, barely missing one another then arcing round to charge each other again, forming a figure-of-eight pattern in the air. Occasionally a third bee was involved, and then the patterns became chaotic as they frantically zoomed around, occasionally bashing into one another in mid-air. Sometimes they would lock together in a tangle of legs, fur and wings, and fall to the ground for a moment of two before resuming the aerial dogfight. One battle I watched went on for nearly four minutes – I began to feel tired just watching them. Perhaps it was a test of endurance? Inevitably one bee would land first, looking exhausted, and if it returned to its perch it was then attacked by the bee remaining in the air, which would swoop down on top of it, forcing it to return to the fray. The usual outcome was that all the males would eventually retire to their perches, but once in a while one seemed to be driven off, perhaps unable to keep up with the pace, or perhaps needing to top up its energy reserves with some nectar. When this happened his perch was always stolen.

I experimented with catching bees and keeping them in a pot for a while. If they were from a group of bees, it was usually not long before another bee would come along and occupy their spot. It is probably anthropomorphising excessively, but it always looked as if the intruder was particularly pleased with himself but still nervous, expecting to be attacked at any second for its temerity in stealing another's favourite perch. Some perches seemed to be especially popular – if I removed a second bee then a third would soon turn up and take it over. So what were these bees up to?

Interestingly, there is a direct parallel with another creature that lives in these same forests. The cock-of-the-rock is a bizarre and spectacular bright red and black bird with extraordinary crests resembling those on a Trojan helmet – it is the emblem of the Santa Lucia Reserve. Every September, during the mating season, males come together in the morning in what is known as a lek – a gathering of males to which females come to select their mate. The lek at Santa Lucia is a stiff two-hour walk from the lodge, tucked deep in the forest in a steep-sided valley. The lek takes place just after first light, for about half an hour, so to observe it one must get up at 4 a.m. and do the two-hour walk by torchlight. I couldn't resist giving it a go, and it was worth the effort to see such a peculiar and fascinating sight: a group of five males performing a head-bobbing dance while fluttering their wings and making hoarse, throaty squawks. Apparently, some days there can be twenty or more birds there, so I caught it on a quiet day. The idea is that a female looking for a mate just has to come along and watch – all the males from roundabout are gathered here, strutting their stuff, and she can pick the most attractive as her mate. By doing so, her sons will hopefully inherit the same sexy characteristics and be desirable mates, so furthering the family line. The males don't contribute anything to looking after the offspring, however, for after a quick copulation the chosen male returns to the lek. If he is particularly desirable, he may get to mate many times, while his less impressive rivals may not mate at all. While I was there I didn't see any females visit the lek; I guess that they turn up very rarely, so that most mornings the antics of the males are to no avail.

It isn't just cock-of-the-rocks that lek, for in the UK black grouse do it, and the great bustards on Salisbury Plain. The clouds of black flies that one often sees in summer, dancing together in sheltered places near streams, are probably doing something similar. A few species of antelope, wasp and fish also lek. It is easy to see the benefits for the females, for they get to select the pick of the crop, and the system is great for the most handsome males, but it must be a frustrating business for the less fortunate males who are forced into a competition that they cannot win.

Presumably my bumblebees were doing something similar, but in a less aggregated and more aggressive way than the cock-of-the-rocks – instead of dozens of males at one hotspot they were in small clusters, sometimes on their own, and instead of just displaying, the males indulged in aerial combat. So far as we know, male bumblebees only really have one function in life – mating – so their behaviour ought to relate to finding a mate in some way. It is very hard to conceive what else it might be for. I guess that females must come along occasionally, like female cock-of-the-rocks (which perhaps should be called hen-of-the-rocks?), possibly attracted by the pheromone, or perhaps attracted by the same mysterious qualities that the males see in their perching locations. If she finds a single male guarding his lonely post, would she mate with him, or reject him as a sad loser who can't manage to obtain a perch at one of the prime locations where several males gather? Would she travel on, looking for a group of males so that she can compare them and choose the sexiest? If so, then the lone males are wasting their time, and they should try to join a larger group, like a male cock-of-the-rock. Over time, perhaps that will happen. Maybe these are incipient leks, and in a few thousand years the males will be gathered in even larger groups. Or maybe the females aren't so fussy, or don't need to make a side-by-side comparison before choosing a mate, so that solitary males occasionally get lucky?

We don't know the answer to any of these questions. If I can, I want to go back, to try to understand them better. I never saw a single queen, but presumably virgin queens do turn up occasionally. Perhaps I could film the male aggregations, and so stand a better chance of catching a visit by a queen. I could try pasting lots of pheromones onto leaves to see whether they attract queens, but that would involve getting the pheromone out of the heads of the males, which they almost certainly wouldn't enjoy. I could measure the males, to see if those that occupy the clusters are bigger or stronger than those perched on their own. I could DNA fingerprint the males, to see if the clusters consist of brothers or unrelated males, and I could see if the aggregations of males tend to be in the same place each year. There are so many possibilities.

They may all seem rather trivial – why does it matter what these bees are doing? But then Darwin's studies could have been dismissed as trivial, were it not for the fact that ultimately he came up with perhaps the most important theory that science has yet devised. While his contemporaries were engaged in obviously practical endeavours such as developing new types of steam engine and building the foundations of industrial chemistry, Darwin spent decades watching worms, scrutinising barnacles, getting his children to chase bees, and comparing the shape of the beaks of finches from the Galapagos. How fantastically esoteric! Who could possibly have predicted that this would lead to something as devastatingly profound as the theory of evolution by natural selection? Of course I'm sure I'll never discover anything one-trillionth as important, but we should not dismiss the value of trying to understand the world around us, for who knows what it might reveal?

On our last day in Santa Lucia I went to watch the nearest group of male bumblebees to the lodge. After two weeks of strenuous exercise and limited food I felt incredibly fit at last, though in desperate need of a huge, gravy-filled meat pie. The bees were fidgety and restless as ever, eyeing each other from their perches. I couldn't help but feel a sense of unease. Three thousand kilometres to the south, in the river valleys below Mendoza in Argentina, the European buff-tailed bumblebees are probably still on the move, heading north. Will they make it up here, to these beautiful, pristine mountains? The Ecuadorian cloud forests are full of flowers, and their altitude means that they are not so hot; I wouldn't be surprised if buff-tails could thrive here, along with their European diseases. We must hope that the buff-tails don't make it, or that if they do, *B. hortulanus* proves to be more resilient than *B. dahlbomii*. That is not the only threat. From where I was sitting I had a narrow view southwards from between the forest trees, to the village in the valley below and beyond. The valley is bright green compared to the forest on the mountain slopes. It has been cleared for sugar cane or pasture, and there are some of these patches of bright green scattered across the mountain slopes, where farmers trying to increase their income have begun clearing the steep sides of the mountains. Santa Lucia is now protected, at least as long as the reserve can keep afloat financially, but steady deforestation is occurring all around. Mika's estimates from satellite data suggest that 0.7 per cent of the surrounding forests are lost every year. That may not sound like a huge amount, but it means a little less room for pumas, butterflies and coatis every year, and if it goes on for long enough then there will be nothing left. The large mammals are usually the first to go for they need huge ranges, and as humans move in to farm they come into conflict with them. Hunting is still prevalent, and even jaguars, protected under law, are still commonly shot.

It is a huge challenge to reverse these trends, to reconcile the needs of humans and wildlife, but we have to find a way. Santa Lucia provides one possible model – protecting the forests through tourism and providing facilities for scientific study – but it is not enough. A colleague at Sussex, Jörn Scharlemann, recently calculated what it would cost to set up a network of protected areas that would conserve every endangered bird species in the world. Such a network would, of course, also protect most other endangered species on Earth. The figure – seventy-six billion dollars per year – sounds astronomical, but as he points out, that is just

20 per cent of the global annual spend on fizzy drinks, and less than half of what is paid out each year in bonuses to bankers in Wall Street's investment banks. Put like that, it is not such a big ask. Of course this sum is only the tiniest fraction of the amount we spend on waging wars. We could easily afford to protect cock-of-the-rocks and cockroaches, hummingbirds and hawk moths, bears and bumblebees, if only we chose to. One can of Coke in five – not much of a price to pay to save the world.

CHAPTER SEVEN

Brownfield Rainforests of the Thames Estuary

In a field that is claimed to be wasteland
only fed by the sun and the rain;
it's dotted with a sprinkling of yellow
amongst the wild grasses again.

Lindsay Laurie, from 'Dandelion'

As you will have gathered, I have always been somewhat obsessed with wildlife. From the age of about seven I spent my weekends and summer holidays catching newts and great diving beetles in the local canal, scrambling about in disused quarries for rare orchids, clambering about in an abandoned mill in search of birds' nests, or hunting for butterflies on the wasteground around some nearby gravel pits. This was Shropshire, the county where the Industrial Revolution began, and these places where my friends and I went hunting for beasts were abandoned relics of that industrial past. I never thought about it at the time, but my friends and I hardly ever went hunting butterflies in farmland, as we had long ago learned that there was rarely much interesting wildlife to be found in the bright green pastures of rye grass or the vast monocultures of wheat.

You may wonder why these industrial scars sometimes end up supporting so much wildlife. In some cases the answer is obvious – a canal may have been constructed to transport coal or iron ore 150 years ago, for example, but it is at the end of the day just a lengthy, very thin lake, and so long as it isn't heavily polluted it is bound to be colonised by whirligig beetles, dragonflies, water beetles, kingfishers and a host of other creatures. Many of the ponds that were once a feature of farmland have been filled in, but quite a lot of canals remain (though sadly an awful lot were drained and filled in). Where they survive, canals teem with life, and luckily most that were not destroyed are now looked after and valued as places to fish, cycle along the towpath, birdwatch, or just somewhere to escape to for a bit of peace and quiet, away from the busy roads and hubbub of modern life. But what about quarries, abandoned factories, mine spoil heaps and so on – why on earth should they become valuable reservoirs of wildlife? It is in part simply because they have been abandoned; they are no longer disturbed by man, no pesticides are used, they are not ploughed or cropped. Nature tends to creep back into the most unlikely places if given half a chance. The McLaughlin Reserve I had visited in California is a great example; once a gold mine, now it is a haven for rare bumblebees, rattlesnakes and red-legged frogs.

One feature that many abandoned industrial sites have in common is that the soil is poor, if there is any soil at all. This may seem counter-intuitive, but much of our wild flora is adapted to low-fertility soils; the widespread use of cheap artificial fertilisers has rendered most farmland too fertile for them to thrive. Next time you find yourself walking along the edge of a field next to a hedge, look at what plant species you can see in the hedge margin. You will rarely see orchids, scabious, harebells or cowslips – instead, wherever you are in Britain, you are likely to see a profusion of nettles, docks and cow parsley sprouting like triffids, for these are among the few species that thrive in high-nitrate soils. In contrast, amongst the broken concrete, spoil heaps, gravel and bedrock of former industrial sites, there are usually few nutrients. Here, many species of plant and insect have found a suitable home where they can thrive in peace.

The higgledy-piggledy nature of former industrial sites can also offer many different niches for plants and animals to exploit. A quarry is likely to have sun-traps, south-facing sheltered corners where warmth-loving insects and spiders may thrive, and damp, shady corners that the sun never reaches where liverworts and mosses can prosper. There will be cliff ledges for birds to nest on safely, and where pinks and stonecrops might gain a foothold amongst the cracks in the rock. In the quarry bottom water may collect in shallow pools and form marshy areas where amphibians can breed, and damp-loving orchids may spring up. None of this was intended by the workmen who blasted or chipped a hole in the ground all those years ago – at the time, it may have seemed that they were making a terrible mess of the countryside, and local folk may have complained at the dust and noise. The hole the quarrymen made would usually have been abandoned simply because it was no further use and there was no easy way to fill it in, yet in doing so they accidentally created a haven for wildlife.

Unfortunately, former industrial areas tend to be classified as 'brownfield' sites. The very name is unattractive – 'brownfield' sounds dirty, and in the popular imagination the word conjures up images of eyesores: polluted, rusting factories and slag heaps, carbuncles in urgent need of remediation followed by redevelopment. What is more, brownfield sites are usually close to or in cities, where land prices are sky-high and developers endlessly on the lookout for land to satisfy our insatiable demand for new housing,[36] new out-of-town shopping centres, and so on.

[36] The usual argument that is trotted out is that we need thousands more houses to make housing more affordable so that young people

On the face of it, developing brownfield sites makes perfect sense, if you accept that we need new developments at all. The obvious alternative is to develop greenfield sites, and of course nobody wants that. Greenfield is usually regarded as sacrosanct – Britain's green and pleasant land – and politicians regularly pledge to protect it (though of course plenty of greenfield sites are developed every year). But let's reflect on this paradox for a moment.

Greenfield site usually means farmland – fields of cereals, oilseed rape or improved, bright green pasture. These fields generally harbour almost no wildlife – they are close to perfect monocultures, without weeds or insects or birds. Go and stand in a field of wheat in high summer and the chances are that you will see nothing but wheat, hear no sound but the wind. There will be no buzz of insects, and probably no twitter of a skylark high above. If there are lots of thick hedges, wide field margins and unfarmed corners then these might support a reasonable number of common plants, birds, insects, and so on, but on the whole most farmland is a desert for wildlife. Ironically, building houses on farmland will result in a significant net increase in the numbers of most types of wildlife – bumblebee populations, for example, seem to be much stronger in gardens than on farmland. When we placed young bumblebee nests in gardens and on farmland and monitored how they grew, the results were clear as day: they tend to do much better in gardens. Honeybee keepers tend to get much higher yields in urban areas, so much so that there has been a huge surge

---

can get on the property ladder, but does anyone really believe that building 100,000 new homes will actually reduce house prices? I suspect that the real motivation is that huge housing developments create massive profits for the powerful companies involved in house building, road construction, and so on.

of enthusiasm for beekeeping even in central London, and there are now many hundreds of honeybee hives perched on top of office blocks and hotels. City dwellers tend to think that wildlife lives in the countryside, but oddly enough there is often more in town these days. That is a pretty sad indictment of modern farming, but it also shows the potential of gardens as urban nature reserves, an opportunity that we should make more of if the area of suburbia is to continue to increase, as it surely will.

Of course there are other reasons not to build on farmland, not least the rather obvious one that we need food. Governments often talk about the desirability of being more self-sufficient in food production, so that we do not depend too heavily on imports from abroad, but then they approve the building of entire new towns in rural locations, more often than not on productive farmland. My point is that we should not automatically assume that brownfield development is preferable to greenfield. Each site needs considering on its merits, there being many conflicting pressures on the land as human populations continue to grow.

I suspect that some readers might think that I've lost the plot on this one, for even suggesting that we might sometimes prefer to build houses on a farmer's green field rather than on a grubby former industrial site. Let me give you an example. But first, let me ask you a question: which site in Britain do you think has the most known species overall, the highest 'biodiversity'? You'll never guess, so I'll tell you – it is Windsor Great Park, a huge area of 3,500 acres containing very ancient oak woodland, lakes and grasslands. Being close to London, it has also been intensely surveyed by biologists of all inclinations for centuries. The list of species known from Windsor Great Park is likely to be somewhere near complete, or at least as near completion as for any site on Earth. Over 2,000 species of beetle alone have been found there, a quite staggering number – who knew that there were this many species in the whole of Britain, let alone living in one place? Now, which site in Britain do you think has the highest density of species, the most per square foot? In truth no one can say with certainty as many places have not been thoroughly searched, but a top contender is a place which could not be more different from Windsor Great Park, though it too is close to London. I had heard much about this place, not least because it has become one of the latest battlegrounds for British conservationists, with its champions coining the term 'Britain's rainforest' to draw attention to the extraordinary biodiversity that the site contains. It had also come to my attention as it was said to harbour populations of some of Britain's rarest bumblebees. In July 2015, I thought that it was finally time to pay it a visit.

My trip got off to an inauspicious start. The directions I had been given told me to head for St Clement's Church in West Thurrock, East London, on the north bank of the Thames Estuary. The area for miles around is one extended industrial estate: roads, roundabouts, concrete hardstanding, huge lorry parks, steel-clad factory buildings and warehouses with graffiti on the walls, litter everywhere. Similar areas can be found sprawling across great tracts of the modern world, especially on the edges of our major cities, and for me they exemplify what an awful shitty mess we are making of the world. Here and there were scraps of vegetation, buddleia, thistles and chicory sprouting on the few bare patches of soil, but it could not have looked much less promising.

My cheap satnav was not up to the task of dealing with the warren of roads and roundabouts and it took me half an hour of going round in circles in this dispiriting landscape before I eventually found the church. It turned out to be a rather lovely old flint-faced building in typical East Anglian style, once situated in rural Essex but long ago swallowed up by the growing urban sprawl. It now crouches incongruously in the shadow of some kind of industrial chemical plant, massive blocky steel buildings painted bright red, adorned with a tangled mass of steel pipes and silver chimneys spouting fumes.

Feeling somewhat dispirited and oppressed by the industrial landscape, I parked and walked south from the church along an overgrown litter-strewn path through thickets of bramble and buddleia that were doing their best to hide the sea of discarded rubbish. It was a very warm, humid day and it was stifling along the narrow track, so it was a relief when after about half a kilometre I came out on to the top of the concrete sea wall overlooking the muddy tidal estuary of the Thames. A path ran east and west along the shore. Judging by the abundant dog faeces, graffiti, discarded beer cans and broken bottles, this was clearly a popular route for dog walkers and an evening hangout for teenagers with nowhere better to go. On the mudflats, a lone curlew took flight, its haunting call seeming out of place in this urban mess.

I turned right, heading west along the sea wall. If my directions were correct, my destination should appear on my right after another kilometre or so. I was looking for West Thurrock Lagoons, a thirty-acre triangle of land which on Google Earth had appeared bright green amongst the concrete sprawl. Part of the site of the old West Thurrock coal-fired power station until the early Nineties, it had been used as a dumping ground for the huge quantities of pulverised fuel ash produced by the station. The power station itself has since been demolished, but the field of ash had been left to its own devices for twenty years or so, and it was this patch of unpromising land that was said to have become a hotspot for wildlife.

I walked westwards towards the Queen Elizabeth II Bridge over the Thames at Dartford, its looming arch jammed with traffic. I was far enough away that the rumble of engines was barely audible, and it was surprisingly peaceful. Powder-blue flowers of chicory lined my path, from which red-tailed bumblebees were busy gathering nectar. Amongst them were quite a few female *Dasypoda hirtipes*, one of our more spectacular solitary bee species. They are sometimes known as pantaloon bees since the females appear to be wearing enormous golden trousers of which a gaucho would be proud. In fact, the hind legs are actually clothed in a bush of long, golden hairs which they use to collect pollen. Bees have evolved an eclectic variety of ways of carrying pollen. The most familiar are the 'baskets' used by bumblebees and honeybees; these comprise a shiny, flattened hind leg segment along the margin of which is a line of long, curved bristles. Pollen is packed against the side of the leg in a neat ball, glued together with a bit of sticky nectar. The pantaloon bees use the same hind legs, but simply trap dry pollen amongst the multitude of hairs rather than creating a sticky ball. Yellow-faced bees carry the pollen inside their stomachs and regurgitate it in the nest, while leafcutter bees pack it amongst the hairs on their belly.

As I watched the pantaloon bees fill their pants with pollen, I began to cheer up a bit. After all, the sun was poking through, and there were plenty of flowers and bees. Clumps of ragwort were smothered in insects, tiny black pollen beetles, vivid green swollen-thighed flower beetles, small skipper and gatekeeper butterflies, and hairy orange tachinid flies. Teasel spikes taller than me were providing food for garden bumblebees and common carders. A little further along, a very sturdy newish fence appeared on my right, about eight feet tall and topped with nasty-looking spikes. Every ten yards there were stern signs – 'Strictly no entry', 'Trespassers will be prosecuted', and 'Caution: guard dogs on patrol', which bore pictures of fierce-looking dogs. Behind the fence all I could see was young birch woodland, some trees perhaps thirty feet tall, which effectively screened everything beyond even from the slightly elevated top of the sea wall. This had to be West

Thurrock Lagoons, but it didn't look as if it was going to be a very productive trip. I hadn't sought permission to access the site, having hoped that I might just be able to wander in, but this was looking unlikely.

I walked along the fence line, scouting for a way in, and came upon a large patch of tufted vetch, a handsome, scrambling plant with clusters of tubular purple flowers that is popular with long-tongued bumblebees. And there, busy amongst the flowers, were no less than three brown-banded carder workers, handsome, straw-coloured insects with bands of richer, rusty orange on their abdomens. The brown-banded carder is the commonest of Britain's rare bumblebees, if that makes any sense – rare enough to get me excited, in any case. As with many of our rarer bumblebees it has become mainly a coastal creature, clinging on in dunes, salt marshes, sea cliffs and other places that aren't amenable to farming. Salisbury Plain is one of very few places where it can be found inland. I wasn't entirely surprised to see it here, for the Thames Estuary is one of the last strongholds of the brown-banded carder and also two much rarer species, the shrill carder and the red-shanked carder. Unfortunately this means that these endangered creatures are directly threatened by the urban sprawl of London as it spreads eastwards along the Thames.

I spent half an hour or so trying to get good photographs of the brown-banded carders, and then headed further along the fence. At the westernmost corner, where the fence made an abrupt right-angle turn and headed inland, someone had been kind enough to smash a goodly sized hole. There was a burnt-out Yamaha off-road motorcycle wedged in the gap – it looked a little as if someone had made the hole by ramming the fence with the bike, though if so that would seem like a near-suicidal strategy. However it had happened, I was grateful, and I climbed through, scrambled across a muddy ditch full of shiny, dancing Dolichopodid flies, and pushed my way through a reed bed to emerge into the birch carr woodland beyond. Since I clearly wasn't really supposed to be there, and there was an outside chance of being attacked by vicious guard dogs, it was all starting to feel like a proper adventure.

To be honest, it wasn't quite what I'd expected. There was a well-worn dirt-bike track through the trees, so clearly bikers came in regularly to race around, presumably also without permission. Their wheels had worn deep ruts in the soft ground, revealing a dark greyish sand that I assumed was the fuel ash. Alongside the track I stumbled across a bizarre memorial, a nicely made wooden cross bearing the inscription 'Andrew Darvill, rest in peace', surrounded by a neat array of several hundred Foster's beer cans. It seemed an unlikely burial place – perhaps it was to commemorate the demise of the biker that had rammed his Yamaha into the fence?[37] I followed the trail northwards through the woods: it was pleasant enough, but the accolade of 'England's rainforest' was seeming a little far-fetched, if I'm honest. I could tell that there was a large open area to my right, towards the centre of the fenced area, but whenever I tried to push my way through the scrub to get to it I came to the banks of a long, reed-filled lagoon that I could not easily cross.

After half a kilometre or so it seemed as if I was nearing the northern limit of the site, for I could hear heavy machinery manoeuvring not far ahead, and could see glimpses of rusting shipping containers through the trees. I turned right once more, and found that the lagoon here was shallow and narrow. By pushing through the reeds and clambering across an old plank

[37] I was later to discover that a family had lived there for several years, constructing shelters from bits of corrugated tin sheeting that they scavenged from nearby factories. However, I'm still not clear as to whether someone is actually buried there.

that had thoughtfully been left there, I suddenly found myself waist-deep in a sea of flowers. I had emerged into an open area of perhaps ten acres in extent, dominated by the frothy pink and white flowers of goat's rue, a shrub-sized legume that clearly thrived in the grey ash. It was interspersed with chrome-yellow splashes of St John's wort, the rich purple drooping spikes of buddleia and dense mauve stands of creeping thistle. Botanically, it was not especially rich, and many of the plants such as the goat's rue are non-natives, but the plants were laying on a splendid banquet for a host of bees, hoverflies and other insects. I was pleased to see several *Volucella zonaria*, Britain's biggest hoverfly, which attempts to overcome the fact that it is a large, tasty and defenceless morsel for birds by mimicking our far-from-harmless native hornet.

The whole lot was alive with insects – mainly bumblebees by the thousand – and I spent the rest of the afternoon wading through the flowers, photographing anything noteworthy that I spotted, and trying to find some of the very rare species that live here. I would have loved to have found the distinguished jumping spider, but try as I might it did not reveal itself. This is one of West Thurrock Lagoons' specialities – a creature known only from two sites in the whole of the UK, the other being the Swanscombe Peninsula on the other side of the Thames in Kent. Both are brownfield sites, although Swanscombe will soon be a huge Paramount theme park, intended to rival Disneyland in Paris as one of Europe's biggest tourist attractions – bad news, of course, for the spider, which probably won't appreciate the water flumes and rollercoasters. It is a very handsome spider, furry and patterned in delicate shades of grey, with two pairs of oversized, forward-pointing black eyes with which it eyes up the distance to its prey. Although I could not find a distinguished jumping spider, I did find several five-banded weevil wasps drinking nectar on the thistles, a species not quite as spectacularly rare as the jumping spider but uncommon enough to get excited about. These are pretty little solitary wasps with, oddly enough, four rather than five clear yellow bands on their abdomen. As the name suggests, they are specialist hunters of weevils,[38] paralysing them and storing them in piles in underground tunnels dug in sandy soil, where the hapless weevils are slowly consumed alive by the wasp grub. I've always been fond of weevils, but I managed to avoid holding their gruesome lifestyle against the rather lovely wasps.

It was baking hot as I hunted about, with the sun now at full strength and the dark ashy soil absorbing its heat, becoming almost too hot to touch. Herein may lie one of the secrets of West Thurrock Lagoons, perhaps one reason why there are so many insects here. Many insects are at the northern edge of their range in Britain, struggling to get through their life cycle each year in our damp and temperate climate. For this reason there are many British insects which tend to be found mainly on south-facing slopes, or in heathlands and sand dunes where patches of bare sandy soil become unusually warm in the summer sun. The grey-black ash provides just such a warm microclimate, radiating heat – it may not be a rainforest, but it felt almost tropically warm.

Amongst the flowers were areas of shallow water, with dark sandy banks that reminded me of the volcanic black-sand beaches of Tenerife. The water was bath-temperature and so I kicked off my shoes, rolled up my trousers and had a paddle, though given the past industrial nature of the site I resisted the temptation to

---

[38] Weevils are a family of vegetarian beetles with endearingly elongate and down-curved snouts, which with a little imagination resemble an elephant's trunk. These generally innocuous and inconspicuous little creatures are spectacularly successful, for there are 40,000 known species.

actually have a swim. With the afternoon sun beginning to dip, I sat on a sandbank, my feet in the water, and soaked up the peculiar atmosphere. It was strangely idyllic – that somewhere this pleasant and so rich in wildlife could have been created entirely by accident from dumping industrial waste was mind-boggling. This entire community had sprung out of nowhere in just twenty years, in the middle of a heavily industrialised landscape, with no help whatsoever from man. Presumably most or all of these creatures must also survive in other small scraps of undisturbed or abandoned land not too far away, often perhaps ones that conservationists have yet to discover. In March 2008, for instance, the conservation organisation Buglife had attempted an inventory of no less than 576 brownfield sites in London and the Thames Gateway. Over half of these sites were found to have significant biodiversity potential, though they did not have the manpower to survey them. Nonetheless, it is clear there is likely to be a network of other wildlife-rich sites right in the middle of the biggest city in Britain, though many of them may be hard to find and may not be accessible to the public. Sadly, the Buglife study also estimated that all would be built on by 2028 at the current rate of development.

A few weeks later in mid-August I came back to West Thurrock Lagoons on an authorised visit. I was keen to explore the site again with someone who knew it well, and also to find out more about its history, so I had arranged to meet Sarah Henshall, Buglife's 'Brownfield Officer', at the official site entrance, tucked away behind a huge printworks. Sarah turned out to be one of those rare people who has a bubbly, infectious enthusiasm for her job, and for all things insect-related, and as we rambled around the site pointing out interesting beasties to each other she related the recent history of the site and the battle to save it. On my previous illicit visit I had, it turned out, managed to discover only the most southerly of three sections to the site, albeit the largest, and each section has had a different fate. Sarah's story began in 2005, when an entomologist named Peter Harvey was asked to survey the site to find out what was there, presumably because the owners were thinking of developing it. At the time, the southern part of the site was known to support important populations of some unusual birds, including nesting reed warblers, sedge warblers and bearded tits, and on this basis it had been granted the status of Site of Special Scientific Interest (often known as an SSSI). Little was known about the insects that lived there, although there had been some perfunctory surveys done in 1996 and 2003. What Peter discovered must have left him both flabbered and gasted in equal proportions. In a few visits through the year he notched up 939 species of invertebrate, which when combined with some previous records brought the site total up to no less than 1,243. This included thirty-five 'Red Data Book' species – those on a register of the most highly endangered species that is maintained by the International Union for the Conservation of Nature – plus 116 'Nationally Scarce' and 352 'Nationally Local' species. Amongst the many bees, Peter of course found the brown-banded carders along with red-shanked carders and the rare sea aster bee (*Colletes halophilus*) that likes to dig its burrows in sandy banks in salt marshes, and seems perfectly able to nest and survive in places below the spring high-tide mark, presumably by sealing up the nest entrance so that it is trapped in a pocket of air. Despite this clever trick, this bee has suffered enormously from reclamation of these marshes over recent decades, removing both nesting sites and the sea aster on which it heavily depends for food.

Peter made all sorts of other unexpected finds, including the salt marsh short-spur beetle (*Anisodactylus poeciloides*), the hump-backed red ant (*Myrmica bessarabica*) and the fancy-legged fly (*Campsicnemus magius*). There was even a tiny fly that lives in reed beds, *Homalura tarsata*, which had never been recorded in Britain before. Of course these are all spectacularly obscure species – almost nobody has ever heard of them, and we know next to nothing about what these creatures do in their daily lives – but it is rather wonderful that they exist nonetheless.

The long and the short of it is that Peter found a truly remarkable number of creatures, especially when you consider that the entire site covers just thirty acres, roughly the same as a single average-sized arable field. And that was without counting the numerous plants, mammals, birds and amphibians, or doing a thorough moth survey with light traps, which would be sure to add a couple of hundred more. This was what triggered the claim that this was Britain's rainforest, in the sense that true rainforests are famous for the huge numbers of species that they contain. The West Thurrock Lagoons was the nearest we had, a treasure trove of weird and wonderful insects and spiders.

As fate would have it, just after the extraordinary value of this site was discovered, the Royal Mail submitted a planning application for a vast warehouse and lorry park which would obliterate all of this diversity. Naturally enough, Buglife took the decision to oppose the development. Their first port of call was Natural England, the government body responsible for protecting biodiversity. Buglife appealed to Natural England, making the persuasive argument that if the most species-rich site in Britain could not be protected by our environmental legislation then nothing was safe. They asked for the whole site to be declared an SSSI, which would have given the area some protection, but Natural England refused to do so. Instead, after some negotiations they accepted a compromise plan whereby the site would be chopped in half. The northerly section would be developed and the southern section, the bit that I had previously explored, would be protected, along with improving a small patch of wetland for the rare birds by digging it up and inserting a pond-liner to retain water. So far as conserving insects was concerned this was a pretty disastrous proposal, for the northernmost section contained the best insect habitat in the form of a fantastic meadow rich in clovers and vetches, where most of the brown-banded carders were to be found.

Buglife launched a petition, which received 2,500 signatures and led to a motion being tabled in the House of Commons. They even managed to get a meeting with the then Prime Minister Tony Blair, but to no avail. Buglife then launched a prolonged and expensive legal challenge to the development, which went through various courts and ended up in the highest court of them all, the House of Lords, where it was rejected. It seems that development needs were regarded as more important than conserving rare species – the nation's need for another lorry park and warehouse, part of our endless race to grow our economy more quickly, took priority over the distinguished jumping spider, the fancy-legged fly and the brown-banded carder bumblebee.

Just when all seemed lost, the Royal Mail scored a spectacular own goal by releasing a pack of commemorative stamps depicting the UK's rarest insects. The wonderful irony was not lost on the campaigners, and Buglife swiftly released their own set of spoof stamps depicting the very insect species that would soon be destroyed by the Royal Mail warehouse. One imagines that there was considerable embarrassment behind the scenes at Royal Mail, and that one or two heads may have rolled. Shortly afterwards they withdrew from the development. Their distribution centre has since been built on another site just a few yards away, one that had no biodiversity value. One might wonder why this solution to everyone's problems could not have been arrived at sooner and with less cost to all concerned.

Sadly, this wasn't the end of it for the site still wasn't protected. Another company moved in and built the huge printworks on top of the northern, and most flower-rich, section of the site. If you read the *Daily Mail*, it was probably printed here. I hadn't known it, but the printworks car park where Sarah and I had met had once, not so long ago, been a sea of flowers and buzzing insects.

Walking south from the car park, we entered what had previously been the central section of the site, an area which now looks like it would be an ideal setting for a post-apocalyptic movie or perhaps an episode of *Dr Who*. Much of the vegetation has been stripped away by motorbikes, leaving acres of bare dark sandy ash, loomed over by rusting electricity pylons; Mad Max would have felt at home. The bike trails I had seen in the woods on the southern section of the site were nothing compared to this. Sarah explained that local dirt-bike riders hold unofficial race meetings here at weekends, even somehow getting a burger van in, and that many hundreds turn up. I'm sure they have a lot of fun – I used to have a dirt bike myself, and know how exciting it is to rip the throttle open and slide the bike through a turn, spraying sand behind you – but they've made a spectacular mess of the plant life. Apparently the security firm have pretty much abandoned attempts to remove the bikers, as they persistently smash down fences to get in, and in any case there is nothing of financial value to protect so it is easier to just let them get on with it.

Even with all this disturbance, this area had a lot of flowers in the parts where the bikers didn't go. Fly ash is quite alkaline, and so it supports flowers that one might normally find on chalky soils – the lacy umbels of wild carrot, spikes of yellow melilot, and the delicate pink stars of centaury. Sarah explained that the distinguished jumping spiders were commonly found here, and that they love to hide under pieces of clinker (bubbly, misshapen lumps of rock, another by-product of the power station furnaces). Clinker has little holes, in which the spiders often hibernate. We spent ages turning bits over, but to my disappointment found no jumping spiders, distinguished or otherwise.

Much of this central section of the site is now owned by the National Grid, who plan to build two new pylons on it. This will no doubt be staggeringly ugly, but given that there are already pylons and chemical plants on view in all directions it hardly matters from an aesthetic point of view. It may ultimately not be such a terrible thing for the wildlife, for it will ensure that no other development can take place, and once the pylons have been built one imagines that the National Grid might take measures to prevent the bikers from racing about beneath them, which would pose obvious safety issues. If so, the land beneath the pylons will probably recover and provide a quiet place where flowers and bees may once again thrive, and perhaps where the distinguished jumping spider can continue to lurk.

In exchange for these sacrifices, the southernmost section of the site is now safe from development. Despite the huge number of rare insects, Natural England still seem more interested in the wading birds that visit the site, and much effort has gone in to clearing the encroaching birches and installing an expensive tidal exchange system to help retain open areas of water. The lagoon area where I had been paddling earlier in the summer was of course not intended for errant entomologists. Although safe from development, it still poses challenges in the future. The trees will need to be kept under control if the sunny, flower-filled meadow areas where most of the insects reside are to be retained. The goat's rue is threatening to squeeze out all of the native plants, and perhaps will need controlling in some way. Occasional visits by bikers to this section may actually do good by creating bare patches and a bit of disturbance, but if the bikers are evicted from the central section by the National Grid then they may turn up en masse, and that would be a disaster. Buglife would love to see this area open to the public, so that families could come to see the wonderful flowers and insect life, but motorbikes and frolicking children seem like a bad combination. The future is far from certain, but at least for the moment this site seems to be more or less safe.

In 2009 Buglife received the *Observer* 'Ethical Award' for its efforts to save the West Thurrock Lagoons site, and their battle through the courts certainly helped to raise awareness as to how weak our environmental legislation is, particularly when it comes to rare insects. Yet the whole affair can hardly be seen as a resounding success for conservation, for a big chunk of the original site has been destroyed, and having now driven around West Thurrock rather more than I would have liked there would seem to be plenty of other derelict industrial sites that could have been developed instead. Fortunately there is at least one brownfield site conservation story that did end pretty well, just a few miles east along the Thames on Canvey Island and it was here that Sarah and I went next.

Canvey Island is an odd place; just a stone's throw east of London, yet it feels remote, bleak and windswept. It comprises about eighteen square kilometres of former salt marsh, none of it more than a metre or two above sea level, and separated from the rest of Essex by just a few narrow creeks which weakly justify its island status. The south-eastern end of the island became a popular seaside holiday destination in the early twentieth century, but as with almost all British seaside towns the tourists dried up when cheap holidays abroad became popular, and now the beach-front amusement arcades, nightclubs and caravan parks look rather tired. Canvey Wick[39] lies on the west of the island, and though just a

[39] Wick is an old Saxon word for a shed in which cheese was made and matured. There was presumably once a settlement at Canvey Wick, or at the very least an active cheesemaker, but now the name refers only to the brownfield site which has no human inhabitants.

mile or so from this urban sprawl, and in sight of an out-of-town Morrisons supermarket and drive-through McDonald's, it feels like another world.

Canvey Island suffered terrible flooding in the winter of 1953, with fifty-eight residents drowned, and shortly afterwards a concrete sea wall was constructed to keep the sea at bay. Until that time Canvey Wick would have been tidal salt marsh, subject to regular flooding, and the initial construction of the sea wall would have been very damaging to the local ecology – coastal mudflats and marshes are very rich habitats for small, mud-dwelling creatures such as worms and molluscs, which support huge populations of wading birds, and all this would have been destroyed by preventing flooding at high tide. The salty, desiccated land would initially have been of little value as farmland, and instead it was used as a dumping ground for sand and silt being dredged out of the Thames to keep shipping lanes open – up to six metres' depth of silt was dumped onto the land. Later, in the 1970s, it was earmarked to become an oil refinery, and development began. Dozens of huge tarmac circles were constructed, intended to form the foundations of the huge circular oil tanks, along with concrete roads to link them up, and even street lighting. Then the oil price fell and the project was abandoned. Canvey Wick was little used for decades, other than as a convenient spot to fly-tip or dump and burn a stolen car, until 2002 when plans were drawn up for its development. Just as with West Thurrock Lagoons, over the intervening twenty-five years since the refinery plans were abandoned this unpromising patch of dumped mud – scattered with burnt-out cars, rusting supermarket trolleys and old mattresses – had developed a remarkable fauna. This included five of Britain's rarest bumblebees, and at least 300 species of moths alone. Thirty of the invertebrates found here are listed in the UK's 'Red Data Book' species list.

Once again, the site was split into three different parts, owned by three different organisations, and the initial development was for the northern eighty acres or so. This chunk had just been acquired by a quango named the East England Development Agency, which proposed to build an upmarket car showroom and large parking lot. To be honest, it doesn't look like a great spot for a car showroom – there isn't much passing traffic and Canvey doesn't appear to be the most affluent of places – but then I'm no expert on such things. Luckily for the bees, however, the proposal ran into an obstacle – a small, four-legged obstacle of the great crested newt variety. For reasons that I have never fathomed, great crested newts enjoy a highly protected status out of all proportion to their rarity (they aren't especially rare at all). Please don't misunderstand me, for I absolutely love great crested newts, and would rather that all wildlife gained their level of protection rather than that it be taken away from the newts. The law of our land states that great crested newts cannot be harmed, so the development was put on hold while the newts were relocated to a new home. This is common practice with new developments – not just newts, but common lizards and slow-worms are also very frequently translocated. Such translocations are amongst the most spectacularly pointless of gestures, a sop to keep conservationists happy and allow developers to say that they have mitigated the damage of their latest project. You may wonder why I say this – surely it is a good thing to save newts from being buried under concrete – so let me explain why in a small digression.

It is generally the case that organisms live at somewhere near the 'carrying capacity' of their environment. Imagine a hypothetical population of deer living on an island. The island is big enough to grow enough grass to support one hundred deer. If there are fewer than one hundred deer, the animals will have a surplus of food, life will be good, mothers will tend to have healthy calves which will tend to survive and so the population will tend to grow. If there are more than a hundred deer, food will tend to be in short supply, the animals will become underweight, weak and susceptible to disease, and few calves will be born. In these circumstances, the population will tend to fall. Population biologists talk about 'density-dependence', by which they mean changes in birth and death rates that tend to push a population towards the carrying capacity. To use another scientific term, the population is in 'stable equilibrium'. Of course the population will wobble around a bit – some years may be warm and wet, causing the grass to grow more than usual, and the deer population to climb, but at some point it will fall back when conditions return to normal. Only if the island is fundamentally altered – for example by building an airport – will the carrying capacity of the island for deer change.

Okay, but what on earth has this got to do with newts or slow-worms? Well, in a translocation, one has to find somewhere to move the animals to. Generally, the answer is the nearest place that already has newts or slow-worms – in this case, the parts of Canvey Wick that weren't going to become a car showroom. The first step in such a translocation is to build a newt-proof barrier around the development site, so that no new newts can arrive, and so that any newts that are removed cannot return home (I have no idea if newts actually do this, but I rather fancy trying some homing experiments to find out). Usually an unsightly strip of thick plastic sheeting is used, dug into the ground and attached to wooden posts at intervals, but the East England Development Agency must have had money to burn and they went for the deluxe version – a shiny galvanised metal wall which is there to this day. Once the wall was erected the newts could be evicted by netting them in the ponds in spring when they come to breed, by placing out refuges, sheets of timber or tin or old carpet tiles which they tend to congregate beneath, or by using pitfall traps dug into the ground. They were then released in the adjacent section of Canvey Wick.

I imagine that by now you have spotted the flaw in this plan. The place where they were released already had a resident population of great crested newts (of course they had all been part of the same population until the tin wall was built). By clearing the northern section of the site of newts, the entire population was squeezed into an area of half the size. It does not take a genius to work out what will happen next – the overcrowded newt population will fall. If 500 newts were moved, then on average 500 newts will die, one way or another, and probably in a slow and lingering way, until the population is reduced to what the environment can support. It might not be the 500 that were moved that die – it would likely be a mix of the resident newts and in-comers – but they would die just the same. Of course exactly the same would happen if the newts were taken to a site far away – if it already had a population of newts, they are likely to already be at or near to the carrying capacity of their environment. The only way that a translocation of this sort works is if one can find a suitable habitat for newts that doesn't already have newts – which is very difficult – or create a brand-new newt habitat somewhere else. To be fair, this latter option can work if the new habitat is created competently, but I have seen plenty of bodge jobs, token efforts that are far too small for the number of animals dumped in to them, or which simply fail to create suitable conditions so that the animals will inevitably die.[40]

[40] Another sad example of a futile translocation is the movement of hedgehogs from the Uists to mainland Scotland. Hedgehogs are not native to these Scottish islands, but were foolishly introduced by a well-meaning gardener who thought that they would eat his slugs. Instead, they quite reasonably preferred to eat the tasty eggs of the

Regardless of all this, the relocation went ahead, and the newts were duly evicted. Once the newts had been removed the site was 'sterilised' by bulldozing the vegetation and then spraying it all with herbicide, destroying a huge chunk of wonderful, flower-rich grassland. The developers were presumably keen to prevent any wildlife of value being discovered on the site that might then pose another obstacle to development, so they adopted a scorched-earth policy.

It was at this point that Buglife came cantering over the horizon, doing their best impression of the US cavalry. Buglife was at the time a newly formed organisation which had been created to fill a glaring gap in the coverage of wildlife charities in the UK. There had long been charities devoted to popular wildlife groups such as birds, mammals and plants, but aside from Butterfly Conservation there was nobody to champion the cause of any of our wild invertebrates (insects, spiders, slugs and snails, worms, and so on), which comprise the majority of the creatures that live in the UK (and elsewhere on Earth). It is easy to see why these organisms had no organisation devoted to their well-being – who would join

---

many rare, ground-nesting birds that live in the Outer Hebrides, and they proliferated at the birds' expense. In understandable preference to culling them, a hugely expensive attempt to catch and relocate them to the mainland has been undertaken – thus far moving 1,600 hogs. This is well-meant work being undertaken by wonderful people, but what happens to those hedgehogs? They are being released into main-land Britain where the hedgehog population has been in free-fall for years. The causes of the decline aren't certain – roadkill and the effects of pesticide on their prey seem likely culprits – but the harsh truth is that the relocated hedgehogs face a pretty bleak future, and one might question whether the money this exercise has cost couldn't have been more wisely spent.

the Earwig Preservation Trust or the Slug Appreciation Society? These creatures don't even have a good collective name – 'invertebrates' sounds very technical and off-putting, and rather vague as it refers to anything without a spine, but then what else can one call them? Creepy-crawlies might be the most familiar word, but that doesn't make them sound very glamorous. The name 'Buglife' seems like a reasonable compromise, although pedantic entomologists might point out that, technically speaking, bugs are one particular group of insects, the aphids, froghoppers, shield bugs and their allies.

Canvey Wick did not have rare birds to get Natural England excited: even more so than West Thurrock Lagoons, its value was in the invertebrates. These include such wonderfully obscure species as the Canvey Island ground beetle (*Scybalicus oblongisculus*), a beast that had not been recorded in Britain for one hundred years until rediscovered at this site, and the scarce emerald damselfly, as the name suggests a fabulous metallic green creature which floats through the air on its filmy wings like an animated jewel. If anyone was going to fight for this site it had to be Buglife, and luckily for the Canvey Island ground beetle and its six-legged chums they stepped in to the breach. One might imagine this would be a tricky campaign to sell to the public – it was the first time that anyone had attempted to protect an ugly, abandoned industrial site strewn with rubbish for the sake of a collection of tiny creatures that almost nobody had ever heard of. However, with some chivvying from Buglife, the *Guardian*'s Environment Editor John Vidal recognised the importance of this site and published a major article about it in May 2003 which caught the public's imagination. More surveys were done of the creatures present, with the current list now exceeding that for West Thurrock Lagoons at over 1,400 species, including many great rarities. Although the northern part of the site had been destroyed, the remaining area of about 200 acres was awarded the status of Site of Special Scientific Interest by Natural England in 2005, providing it with a degree of protection.

Sadly, this hasn't put an end to plans for development. The surviving parts of the site are currently split in ownership – the eastern third is owned by the Land Trust, a charitable organisation, but the remaining two-thirds of the site are owned by Morrisons supermarket chain, and they have aspirations to build on it. They have reasons to be optimistic, for despite the SSSI status the local planning authority recently approved the building of a new dual carriageway straight through the corner of the section owned by the Land Trust. The road has now been built, at a reported cost of £18 million, but has been dubbed by locals the 'Road to Nowhere' since it serves no clear purpose. If part of an SSSI can be destroyed to build a road that doesn't go anywhere one can see why Morrisons might think that they are in with a chance.

Aside from the new dual carriageway, however, little has changed at Canvey Wick over the last ten years. The Land Trust have handed over management of their piece of land to Buglife and the RSPB who now jointly run it as a nature reserve, and the long-term future of this section is secure. In September 2014 it was officially opened for the public to enjoy, complete with interpretive boards and marked trails. It was this section that Sarah showed me around.

We set off along the 'Orchid Trail', though being late August it was too far on in the year for any orchids to be in flower. The track made good use of the concrete roads helpfully installed thirty years earlier to access the never-completed oil tanks, and the weathered concrete, part overgrown and with flowers sprouting from the cracks, was more attractive than you might imagine. Even more so than at West Thurrock Lagoons, it was quickly clear that it was the diversity of habitats available that had allowed the site to become so rich in wildlife. Within one hundred paces one could walk through mixed deciduous woodland, scrub, marsh, sandy heathland and dunes and dry chalky grassland. The underlying substrate seems to vary greatly, from coarse sand to fine silts, with the later building works introducing patches of rubble, concrete and rock. In the sandier, disturbed areas I saw clumps of viper's bugloss, and my hopes of encountering some rare bees immediately rose. Even the round tarmac circles prepared for the oil tanks added diversity, for they were slightly raised above the surrounding land, so that drought-tolerant plants such as narrow-leaved ragwort can cling to their edges, their roots slowly tearing the tarmac apart. Although it was not an especially warm day the black tarmac radiated heat, and comma and peacock butterflies were warming themselves by basking on it. In no time at all I spotted a shrill carder, a worker collecting pollen on a clump of red bartsia growing in a sandy hollow. As we crouched down to get a better look Sarah explained the challenges that Canvey Wick presents.

Even though this chunk is now legally protected, it does not mean that it will always remain as rich in wildlife. Left alone it will all become woodland, which might be great for some woodland birds and plants but would almost certainly result in the disappearance of most of the rare insects and many of the plants such as the orchids. The obvious solution is to cut down some of the trees, but local residents object strongly to this, understandably seeing chainsaws as harming nature rather than encouraging it. As Sarah pointed out, there would not naturally be trees on Canvey Island – trees generally won't grow on salt marsh – but this argument didn't convince the locals and in the end a compromise was reached whereby a shelter belt of trees was retained to maintain the appearance of woodland from a distance. The shelter belt acts as a source of seedlings, however, as one birch tree can produce seventeen million seeds per year – so the battle to keep areas open will be endless. The management team also want to create bare patches of ground by scraping off the surface soil, so providing places for mining bees to burrow and for the seeds of annual plants to sprout. They want to add hummocks and hollows in the process, providing a range of different micro-climates for different plants and animals, but this again has been fought by local people who don't want to see their favourite dog-walking route torn up by bulldozers.

More so than most nature reserves, Canvey Wick seems to draw an eclectic mix of folk. Off-road motorcyclists can be a problem, though thankfully they are clearly not as numerous as at West Thurrock Lagoons. Home-made rocket makers have used the site, for the circular tarmac pads make perfect launch pads, the envy of Cape Canaveral, though there is an obvious associated fire risk and parts of the site have been ravaged by fires. Stargazing societies visit at night, the area being just about far enough from London that the light pollution isn't too bad. A small marijuana plantation has been discovered amongst the bushes. The police have been called to illegal raves. Husky races have taken place. Falconers have been caught flying their Harris hawks, to the considerable consternation of RSPB. All of this must be an ab-solute nightmare to manage (with the exception of the stargazers, who sound like a harmless enough bunch). Yet in a way, I couldn't help but think that it was rather wonderful that this place could be used by so many different people, for such diverse reasons, and still be fantastic for wildlife. Indeed, to some extent these visitors may have helped to create the conditions for wildlife to thrive. The minor bald patches created by the odd motorbike perhaps do more good than harm, and the fires caused by a cas-ually dropped cigarette, a stray rocket or a joy-rider burning a car would have helped to keep the trees from taking over, creating open spaces for butterflies to bask and solitary wasps and bees to excavate their nests.

The orchid trail took us on a loop, passing by the near-empty dual carriageway, and back towards the entrance. Now I understood the layout of the site, it was clear that the path back to the car park skirted along the edge of the northern section of Canvey Wick, the part destroyed by the East of England Development Company before any of the site was declared an SSSI. Having read about the history of Canvey Wick and the destruction of this section I was surprised to see that, rather than bare sterile earth, the whole area was a carpet of wildflowers. The metal newt fence is still there, but clearly the attempts to sterilise the site weren't as effective as had been intended. By clearing the vegetation they created lots of bare earth, and as we have already seen, nature will quickly creep back if given half a chance. Sarah wasn't sure what had happened to the proposed development – the East of England Development Company has disbanded in 2012, so presumably the plans have been forgotten for the moment. We stepped over the knee-high fence and waded through red clover, bird's-foot trefoil and creeping thistles. Although it was late in the year and leaden clouds threatened rain, there were plenty of bees, including perhaps a dozen brown-banded carders and a couple of shrill carders. This area actually had far more flowers than the nature reserve, a result of the recent clearance, and it was obviously an important area for the bees. Sadly this section has no legal protection and it is presumably only a matter of time before somebody revives the plans for a car showroom, or decides that Canvey Island needs another business park. At present only one quarter of Canvey Wick is secure. Would it be too much to hope that the whole 280-acre site could be handed over for Buglife and RSPB to look after? Some of the creatures that live here are spectacularly rare. The shrill carder bee has perhaps six populations left in the whole of the UK after seventy years of rapid decline in the face of agricultural intensification. The Canvey Island ground beetle only lives here and at West Thurrock, so far as we know. Can't we leave them all in peace (except for the disturbance from an occasional husky racer dashing past, and the sporadic bang of rocket launches)? Wouldn't it be wonderful if Morrisons donated their land to show that they care about the environment – how about creating the 'Morrisons bug reserve'?

Although parts of Canvey Wick and West Thurrock Lagoons have been saved, there is no doubt that many other brownfield sites that had become rich in wildlife have already been destroyed. Endangered species turn up in all sorts of unlikely places, if you look hard enough. Just as the Canvey Island ground beetle was thought extinct, so for instance the streaked bombardier beetle (*Brachinus sclopeta*) was rediscovered living in a pile of rubble at a site in the London borough of Newham in 2005 having last been seen in 1928 (at Beachy Head in East Sussex). Regrettably, the particular rubble pile they had been hanging out in was scheduled for a housing development, which has since gone ahead – an unfortunate recurring feature of these stories is that rare species are often only discovered when they are about to be destroyed. Sixty-one of these precious beetles were hastily relocated to a specially created pile of crushed bricks and concrete nearby, but whether their new home will suit them we do not know – they are clearly pretty fussy,[41] and our knowledge of this species' requirements is necessarily scant, so the odds are against them.

[41] Bombardier beetles gain their name from their preposterously unlikely defence mechanism. Their abdomen contains a large chamber into which they secrete hydroquinones and hydrogen peroxide, both highly reactive and noxious chemicals. This chamber links to a smaller reaction chamber which contains chemical catalysts. When the beetle

Neither Buglife nor any other wildlife charities have the manpower to identify and then defend every possible patch of abandoned land and rubble pile, so who knows how many distinguished jumping spiders or streaked bombardier beetles may have found themselves concreted into the foundations of a new supermarket? Clearly the only way that such places can be properly protected is if all development sites are thoroughly surveyed beforehand,[42] and then given proper protection if they are found to support significant numbers of rare or important animals or plants. In the current system, neither of these two steps functions particularly effectively. The government seems happy for places identified as being of Special Scientific Interest, such as Canvey Wick, to be sacrificed on the altar of economic growth, even when alternatives are clearly available.

---

is under attack she squirts the noxious chemicals into the reaction chamber where the catalysts cause an explosive reaction, blasting a cloud of boiling benzoquinones out of her back end with an audible pop – definitely not something to try at home. The beetle can twist her abdomen to direct the boiling, foul-smelling chemical at her assailant – I once picked one up and the evil blast it produced scorched the skin on my fingertips brown. I can't help but wonder if once or twice in their evolutionary history these beetles haven't blown themselves up by accident.

[42] The present system involves a cursory survey of the ecology of the site, heavily focused on a handful of seemingly randomly chosen species, notably great crested newts and any species of bat – lovely creatures, but not obviously more in need of protection than many other beasts. These surveys usually pay scant attention to insects and other invertebrates, and are often carried out by people without the necessary specialist knowledge to spot rare and important species if they are present.

I would love to see more appreciation of the amount of wildlife that is living right under our noses, within our major cities, thriving in the most unlikely of places. Brownfield sites can be wonderfully, surprisingly, rich in wildlife. Most are close to or within urban areas, and thus provide a fantastic opportunity for city dwellers to encounter rare animals and plants within walking distance of their homes. I'm not suggesting that all brownfield sites should become nature reserves, for even I would concede that sometimes there might be better uses for them. But we should use some common sense, and pause for thought before calling in the bulldozers. Britain's wildlife is in trouble, and we cannot afford to lose more of the few places left where it is flourishing.

I'd like to think that future generations of children will be able to pond-dip, find beautiful wildflowers, discover bombardier beetles under rocks, and listen to the sound of birdsong and the buzz of bumblebees. Our increasingly urban population has dwindling opportunity to encounter wildlife, and some brownfield sites provide just such green spaces right on our doorstep. Although they were created by man's actions in the past, they are now wild, quite different from our manicured urban parks, places where natural processes are at play, where nature rules.

Living in a crowded country in an increasingly crowded world, such places are incredibly rare, and we should treasure them.

CHAPTER EIGHT

Knepp Castle and the Forgotten Bees

*Feral . . . imagines the lives we no longer lead but might, the
species that no longer exist but could, and the faculties we no longer
engage but should.*

> Robert MacFarlane, interviewing George Monbiot
> about his book *Feral*

It is not every day that one gets an email from a knight who lives
in a castle, with an invitation to be given lunch and a guided tour
of his estate. I couldn't help but be impressed, whilst also feeling
slightly annoyed with myself for my instinctive urge to be overawed
by someone with a title. I was also tremendously excited, for this
wasn't just any old knight with a castle – this was Sir Charles
Burrell, 10th Baronet, owner of the Knepp Estate, a place that I
had heard tremendous things about ever since moving down to
Sussex. Of course I accepted, and a few days later in the spring
of 2013 I found myself driving up to the front of Knepp Castle,
an enormous and rather elegant castellated mansion built for the
Burrell family by John Nash in 1806. There is also a 'proper' castle
about a kilometre away, built by one of William the Conqueror's
knights just after the Battle of Hastings in 1066, but regrettably
the bulk of it was demolished in the 1700s to build the foundations
of a road which eventually became the A24.

I was half expecting a butler in full regalia to answer the door, but instead it was answered by a young guy in rough outdoor gear who laughed at me when I asked for 'Sir Charles'. He told me that Charlie was to be found down the corridor in his study, which turned out to be a vast, oak-panelled room in which the miscellaneous selection of old desks, tables and chairs were strewn with paperwork, documents, old maps, stuffed animals and photographs. Wading through this clutter was Charlie himself, I guessed somewhere around fifty years old, with a boyish shock of unruly hair and a ready smile – he turned out to be the most affable and welcoming of hosts. Ted Green was also there, perhaps Britain's foremost expert on ancient trees and the vast diversity of life they support. He looked pretty ancient himself, weatherbeaten but with a twinkle in his eye and, it turned out, a wicked sense of humour. Over a coffee, Charlie explained how the Knepp project had come about.

He had taken over managing the family estate in 1983 as a young man, fresh out of the Royal Agricultural College in Cirencester. At the time there were half a dozen tenant farmers, and most of the 3,500-acre estate was intensively farmed. The estate has a well-documented history. Much of it was once a deer park, a favoured hunting ground for King John who had 220 hunting greyhounds stationed at the old castle. In the mid-1500s the deer park was abandoned and an active iron industry grew up in the area, leading to the creation of a huge mill pond over a kilometre long to provide power. When this industry didn't last long, the land became predominantly farmland, a mix of pasture and arable, which it remained until the eighteenth century. Fortunately the mill pond survived, another relict of our industrial past which now supports a host of wildfowl and other aquatic life.

Charlie's family came into ownership of the estate in the 1780s, and shortly afterwards had the new castle built, complete with a new, expansive sheep-grazed park surrounding the buildings, in the form of pasture dotted with oaks that have now grown to enormous girth. The rest of the estate remained farmland, and in 1912 Sir Merrik Burrell founded the Knepp herd of Red Poll cattle, a handsome, blood-red dual-purpose breed used for both milk and meat. Through careful breeding he created prize-winning animals, and the estate became strongly associated with the breed, with a herd that grew to contain over 500 cattle by the 1970s. However, the estate was never particularly productive, situated as it is in West Sussex, on heavy Weald clay, a soil not well suited to modern intensive farming. Between the two World Wars half of the land was abandoned as it was simply not profitable to farm it, but during the Second World War the drive to be self-sufficient in food meant that all of the land was brought back into production, including the deer park, with the land ploughed right up to the door of the new castle.

When Charlie took over, little had changed since the Second World War. More or less the entire estate was intensively farmed, although it was barely making a profit as the soil just could not be persuaded to produce high yields of arable crops. To try to make ends meet they farmed the land harder and harder, ploughing right up to the base of the hedges and in tight circles around the great oaks in what had once been the deer park in an attempt to produce as much crop as possible. Even so, shifting subsidy systems and drops in milk prices meant that the farming began to run at a loss. Charlie questioned what they were doing – squeezing every last drop they could from the land for the privilege of losing money didn't seem to make any sense, but after centuries of farming what else were they to do? With great sadness, in the year 2000 they sold off the last of the cattle and abandoned dairy farming altogether.

In 2001, in an attempt to diversify their income, Charlie entered into a Countryside Stewardship Agreement with Defra[43] to restore the deer park around the new castle. Ted Green paid a visit, and was horrified to see the land being ploughed under the mighty oaks, some of which were in a pretty poor state. Old oaks support quite staggering biodiversity – no less than 423 different species of insects and mites have been found living in or on them, eating their leaves, forming galls, sucking their sap, nibbling their roots, and so on. Countless fungi, and a mind-boggling 324 different species of lichen have been found growing upon them (how many of us would have guessed that there were that many types of lichen at all, let alone just on oak trees?). A single oak tree can be a whole world in itself and this continues long after its death, for many of these insects burrow in the dead timber as it slowly decays over hundreds of years. Our ancient forests would have contained great volumes of dead, rotting wood, perhaps as much as 200 cubic metres per hectare, and numerous insects and fungi specialised in feeding upon this resource. In turn, birds such as woodpeckers specialised in feeding on those insects, while other birds and bats would nest in the cavities within the slowly decaying trees. While it is decaying, dead wood also helps to prevent soil erosion, and locks up huge volumes of carbon, while its eventual decay releases nutrients that new trees need to grow. In short, dead trees are wonderful places for wildlife, and an integral part of woodland ecosystems.

It is thus rather depressing to report that modern forestry practice is to remove dead trees as swiftly as possible so that they do not harbour disease, and to avoid the extremely unlikely event that they might fall on a passer-by (how many of us make a habit

---

[43] The Department for the Environment, Farming & Rural Affairs.

of standing around under long-dead trees during windy weather?). In some heavily managed woodland, there can be less than one cubic metre of dead wood per hectare. As a result, many of the creatures that feed on dead wood have become enormously rare – for example, the pine hoverfly, a species that breeds in water-filled rot holes in old pine tree stumps, is now teetering on the edge of extinction in the UK. Similarly the violet click beetle, which breeds in the soggy black mush to be found in the hollow heart of long-dead, well-decayed tree trunks, is now known from just a handful of trees. We have left these poor creatures with nowhere to live in the modern world.

Some of Knepp's oaks were dying or dead, and traditional practice would, of course, be to chop them down and saw up the timber for firewood, but fortunately Ted knew the value of dead wood. He suggested that Charlie leave the dead trees in situ, and plant a few extra ones to eventually replace them. I had seen this when coming up the long drive to the house – the old trees remain, most alive, some part-dead with hollow trunks and craggy leafless branches but some green tufts of fresh leaves, others entirely dead, riddled with woodpecker holes, and slowly shedding their mighty branches to the ground, where they are left to rot. All of them looked magnificent.

As part of the Countryside Stewardship Agreement, the arable fields were returned to pasture, and Charlie introduced a range of grazing animals (fallow deer, Exmoor ponies, old English long-horn cattle and Tamworth pigs), using tough breeds that can survive outdoors with minimal management. Instead of being surrounded by fields of wheat and oilseed rape, Charlie and his family could now look out from their house on herds of grazing animals leading more or less natural lives. They also noticed that insect and bird populations began to increase. As Charlie put it, 'There was a great feeling of ease and space and lack of stress.'

It was this in part that inspired Charlie to make a very bold move – to take the entire estate out of conventional farming and turn it into a 'rewilding' project. The concept of rewilding is argued by some to be a revolution in conservation. Conventional conservation practices in the UK often involve very intensive management to maintain particular, valued habitats. For example, chalk downland will scrub over if not managed, as is happening around the impact zone on Salisbury Plain. The fragments of flower-rich chalk downland that survive, aside from on Salisbury Plain, are almost all in nature reserves such as Box Hill in Surrey, or Castle Hill near Brighton on the South Downs. In the winter teams of volunteers can be seen with axes and saws, labouring to bash back the hawthorn and blackthorn scrub. Grazing animals may be brought in for short periods, or the grasslands cut using a tractor. Invasive weeds may be pulled up or herbicided off. Heathlands are similar, for unless invasive birch trees are removed, our heathlands, much prized for the rare butterflies, birds, reptiles and flowers they support, would quickly turn into forests. These are not natural, wild areas by any means; they are habitats created by the actions of man centuries or millennia ago, and they may be just as intensively managed, in a way, as the cereal fields that surround them.

The term rewilding was first used in 1990 by David Foreman, an American conservationist and environmental activist who was disillusioned by the failure of the conservation movement to make any substantial headway against the tide of habitat loss and extinctions. It refers to a concept of creating large reserves and then allowing nature to decide what happens – to interfere as little as possible, ideally not at all. The best known example in Europe is the Oostvaardersplassen in the Netherlands, a fifty-six-square-kilometre area that was largely open sea until 1968, when it was reclaimed by enclosing it with a dyke. The marshy reserve quickly became a very important wetland for wading birds, but it was threatened by encroachment with willow seedlings which, left unchecked, could have eventually made the area unsuitable for waders. It was then that a Dutch biologist named Frans Vera proposed a controversial solution.

It had long been thought that, before man arrived, much of Europe was covered in dense, closed canopy, 'primeval' forests, an idea that was based on analysis of pollen preserved in peat and lake beds. Frans challenged this idea, arguing that these forests may have been much more open, a patchwork of open glades, scrub and closed forest, maintained by large grazing animals – aurochs, boar, bison, deer, tarpan and so on. Going back even further, 50,000 years ago there would have been rhinoceros and straight-tusked elephants, the latter readily capable of flattening sizeable trees. It has even been argued that the huge trunks and thick, heavily fissured barks of some of our native trees, such as oaks, are an adaptation to protect them from uprooting by these long-extinct creatures. It is quite an attractive idea, for it provides a possible solution to a major ecological conundrum – where did all of our rare grassland flowers, bees and butterflies live before humans came along and cleared the forests? Creatures such as the Adonis blue butterfly, which occurs only on south-facing, sunny chalk downland, could not possibly survive in forests. But perhaps Europe wasn't all dense forests, and the Adonis blues,[44] spider

[44] My former PhD student Georgina Harper came up with an alternative idea. She carried out genetic studies of all of the Adonis blue butterfly populations in the UK, and came to the surprising conclusion that they all had a common ancestor – they were all descended from a single female – that lived perhaps 240 years ago, and was closely related to the Adonis blues of northern France. Perhaps this female was blown across the Channel in a storm, or it might be that this species was deliberately introduced to the UK by a butterfly enthusiast

orchids and so on all lived in these large clearings? Frans persuaded the Dutch authorities that, instead of clearing the willows by hand or with herbicides, perhaps herds of large mammals would do the same job but in a more natural way. Red deer, ponies and cattle were duly introduced, and left to get on with it. The absence of any large predators to control them meant that numbers rapidly rose, and now some animals are shot when they become close to death through starvation. Otherwise management is minimal, and there is currently a resident population of some 2,000 red deer, 800 ponies and about 160 cattle. The animals have certainly helped to keep the willows at bay, and so the area remains an enormously important bird reserve, with spectacular species such as the white-tailed sea eagle in residence. Perhaps most importantly, Oostvaardersplassen and the concept of rewilding have caught the imagination of the Dutch people, not least because the reserve is just ten kilometres from central Amsterdam, so that it is very easy to visit and experience this near-wild area.

Elsewhere, where protected areas are larger, rewilding projects are able to take a further step. The herds of deer, bison, aurochs and ponies that roamed prehistoric Europe were once preyed upon by wolves, lynx and bears, and if one goes back further in time even by lions and hyena. Natural communities all once contained

---

in the 1700s. It just so happens that the timing corresponds with the era when studying and collecting butterflies became a fashionable hobby. Interestingly, Adonis blues were not described as a species in the UK until 1775, long after most other butterflies, and despite the fact that this is a very pretty species which lives in the south where most butterfly collectors were active. Of course it is highly improbable that all of our grassland insects and plant species were brought in by man after we had cleared the forests, but perhaps a few of the more striking ones were.

top predators, until man arrived, drove them away or killed them and took their place. If one wishes to completely restore natural communities and natural ecological processes, then one needs top predators, or so many advocates of rewilding believe. The best known and most successful example of a reintroduction of a top predator is the release of wolves at Yellowstone National Park in the state of Wyoming, USA. Yellowstone was made a National Park in 1872, and covers a vast area totalling nearly 9,000 square kilometres of the eastern Rocky Mountains (to put the size in context, Yellowstone is larger than the counties of Hampshire, East and West Sussex and Surrey all combined). Grey wolves were, of course, a native species in the region, but were nonetheless regarded as undesirable even in a National Park and were deliberately hunted to extinction, both here and across much of their US range. Two pups were killed by park rangers in 1926, and shortly thereafter the wolf seems to have become extinct in the park.

By 1933, there were reports that Yellowstone was being denuded of vegetation by the rapid growth of the elk[45] population in the absence of wolves. The trees had stopped regenerating because any young saplings were being eaten, while the grasslands were being grazed bare. The park service started a long-term programme of shooting the elk, but they didn't manage to kill enough for the state of the ranges to improve. By the 1960s, local hunters who had become used to elk being unnaturally abundant and hence easy to shoot, started complaining that the park rangers were

---

[45] The word elk in North American parlance refers to a species of deer very closely related to the European red deer. We use the word elk to describe a completely different animal, known to North Americans as a moose. In situations like this it quickly becomes clear why Carl Linnaeus's system of standardised Latin names avoids a lot of confusion.

shooting too many. Heaven forbid that they would have to make any effort to stalk the animals before gunning them down. Hunters being a powerful lobby in the US, Congress threatened to pull the funding for Yellowstone if the rangers didn't allow elk numbers back up, even though it was transparently obvious that there were far too many already. Elk numbers were duly allowed to grow again, and the park was stripped almost bare of vegetation.

With the benefit of hindsight, this whole saga seems spectacularly foolish, and the solution to the problem of overgrazing by elk self-evident – reintroduce the wolves. Of course some people suggested this, but it was not until the 1980s that the idea began to be taken seriously. Finally, after decades of argument, in 1995 fourteen Canadian wolves were introduced to large pens at Yellowstone, and after a few months of acclimatisation the gates were opened. The next year a further seventeen were added.

Despite occasional persecution from disgruntled hunters, the wolves thrived. Within just four years there were over a hundred, and the numbers have since fluctuated between eighty and 170. Some have spread beyond the park boundary, where they can legally be shot, but this is a pretty wild part of the world, and many survive – in total, there are thought to be perhaps 250 in the park and surrounding ranges, which may sound like a lot until you reflect upon the size of the area involved. Even if all of these wolves were in the park, it still only amounts to one per thirty-six square kilometres.

Despite this low density, the wolves seem to have had a huge effect on the Yellowstone ecosystem, far beyond simply reducing the elk numbers, which have dropped by about 50 per cent, effects that have been documented in detail by William Ripple, an ecologist from Oregon State University. He argues that the presence of wolves has not just reduced the number of elk but has profoundly changed the behaviour of those that remain. In particular, they tend to avoid steep-sided river valleys, where it is harder for them to see wolves approaching, and instead they stay out in the open. Elk love to browse aspen, cottonwood and willow, and willow grows in damp places along riverbanks. Without wolves the extent of woodland had declined considerably, which had impacted on beavers, which rely heavily on willows for their winter fodder. Beavers had all but disappeared from Yellowstone, but with the wolves back, the willows began to recover and the beavers with them. Beavers themselves are ecosystem engineers – their dams create more wetlands and marshes, encouraging yet more willow to grow, and also providing habitat for amphibians, wading birds, and so on. The diversity of aquatic habitats they create, from deep pools to fast riffles to shallow marshes, also encourage a range of insects and fish species. Their dams also help to store water, reducing bank erosion and flooding downstream during heavy rain – all in all, for a chubby brown rodent with outsized teeth they have a pretty impressive impact.

According to Ripple, the benefits of the wolf reintroduction were seen elsewhere too. Coyotes had thrived in the absence of wolves, but they were driven back to steeper ground where they could escape being depredated, which benefited populations of small mammals and ground-nesting birds that the coyotes had been eating. The remains of elk carcasses left by the wolves provided food for scavengers such as ravens, wolverines, bald eagles, even bears, the latter also benefiting from the revival of berry-bearing shrubs along the riverbanks. The ripples of effects radiating outward from the wolves are often described as a trophic cascade, and it makes a captivating story in which restoring the ecological balance has had profound benefits for wildlife across a huge area. The tale has been repeated often, in magazines, books and nature documentaries, and it has captured the hearts of many.

Some scientists have recently questioned whether the story is quite as neat as described by Bill Ripple. Some attempts to test his ideas have not supported them – for example excluding elk from riverbanks does not necessarily result in increased willow growth, and finding direct evidence that elk are more likely to be depredated if they visit the riverbanks has also proved difficult. Nonetheless, most agree that the wolves and then the beavers have had profound and positive impacts on Yellowstone.

After travelling to Oostvaardersplassen to meet Frans and observe first hand what was happening there, Charlie decided to try something similar at Knepp. In 2004 he took more land out of conventional production, and in 2009 the bulk of the estate went into a 'Higher Level Stewardship' agreement by which he obtains fairly substantial subsidies from Natural England for supporting biodiversity. The land was entirely enclosed with tall deer fencing, unfortunately broken into three blocks because of the roads that run through the estate. Internal fences and gates were then removed, leaving only the old network of hedges. There were already roe deer on site, and to these were added old English longhorn cattle, Tamworth pigs, fallow and red deer, and Exmoor ponies. Charlie would have dearly loved to introduce wild boar, a native species in the UK but hunted to extinction perhaps 700 years ago. Although it is legal to obtain and keep them, they have to be contained within boar-proof fencing which would have been prohibitively costly to erect around the whole estate, so the Tamworths were chosen as the nearest thing. As with the cattle and ponies, Tamworths are an ancient breed that can pretty much look after themselves, and with much longer snouts than other domestic breeds so they are well suited to rooting about for natural food. They are also very resistant to sunburn – it wouldn't be much of a rewilding project if Charlie and his team had to rush out and apply sunscreen to the pigs whenever the sun came out.

Ever since, the animals have been more or less left to look after themselves. The deer population has grown and they are now regularly culled to keep them in check. Likewise, surplus cattle are sent to slaughter, and their meat fetches a premium price in London markets. Otherwise, the animals are left in peace; they are outdoors all year, of course, fending for themselves, giving birth naturally and looking after their young as animals instinctively do. The vegetation is not managed in any way – the land is simply left to develop as it will, with no attempt to keep it clear of trees or to steer it in any way at all.

It was now about ten years on from when this grand experiment had started, and I was desperately excited to see what was actually happening out in the fields. Charlie led us outside to where he had a huge old vehicle parked, a six-wheeled ex-Austrian army open-topped troop carrier, capable of going more or less anywhere. We jumped in the back and off we roared. It felt as if we were off on a proper adventure, a feeling one doesn't often get in Sussex.

What struck me first as we bumped and jolted across the estate was how variable the different fields were. Some were open grassland – yellow with buttercups, and dotted with cowpats and horse dung, though we didn't initially see any large animals. Rabbits scampered away in front of us, and the remains of ones that had been eaten by buzzards – bits of fur and bones, the odd furry foot – were scattered on the grass. When we bounced through a gap in the hedge where once a gate had separated the fields, the next field had an entirely different character. Scrub had moved in – bramble and wild rose, the delicate flushed-pink blooms of the latter just beginning to open. Spiny shrubs – blackthorn and hawthorn – had also colonised, and together these four plant species formed dense defensive clumps, fortresses against the livestock, some of which had grown to be several metres across and a couple of metres tall. We stopped and climbed out for a closer look. Despite their thorns and spines the plants were clearly still grazed, with their outer leaves heavily nibbled and shoots bitten off. Nonetheless, they were clearly growing outwards, albeit slowly. Fascinatingly, from the centre of the larger clumps, just beyond the reach of the grazers, the delicate fronds of ash, oak and birch were protruding. They could never have survived without the protective shield of the thorny shrubs around them. Clearly these fields were slowly turning to woodland, the thorny plants that are able to tolerate grazing ultimately bringing about their own demise, for they will eventually be shaded out by the trees growing through them, though it might take another fifty or one hundred years before anything like a closed canopy can form above them.

When I kicked over a drying cowpat – an antisocial behaviour that entomologists are prone to – I was struck by how many dung beetles and their larvae were burrowing within. A moment's reflection explained why. These cows were not being regularly dosed up with Avermectins, worming drugs that are routinely given to livestock around the world and render their dung toxic to insect life, in turn reducing the abundance of prey for swallows, starlings, and so on. These pats were au naturel, undoctored and hence a fabulous breeding ground for a host of creepy-crawlies – one single dung pat at Knepp was found to be inhabited by no less than twenty-three different species of dung beetle.

Evidence of rootling by the pigs was everywhere – great patches of earth, sometimes half an acre or so in extent, where the sods had been ripped up and scattered. I had seen this in the forests of France and Spain where wild boar abound, but never before in Britain. It looked messy, but in ecological terms this disturbance is a natural process, and probably an important one. The churning of the soil provides bare patches, hillocks and hollows, providing places for plants to germinate and warm, sheltered microclimates for bees and butterflies to bask in. We associate rare arable weeds such as cornflowers and corncockle with the disturbance created by arable farming, where they once thrived amongst the crops until modern herbicides and seed-cleaning methods eradicated them. One might wonder where they lived before the arrival of man – perhaps the answer is that they once depended on the action of wild boar to create patches of bare ground for them?

We jumped back in and clattered through a couple more similar fields, but then the next was completely different – a dense stand of sallow trees, already ten metres high, and packed together so closely that even on foot it was a struggle to squeeze between them. Why was this field so different? Charlie explained that in part it was down to the behaviour of the grazing animals, especially the cows. When first introduced, they had huddled together in the centre of the field, unable to comprehend their sudden freedom. All of their previous lives they had lived in enclosed fields, and they didn't know what to do. It took them weeks to venture into the next field, and months before they had explored the whole estate. Some areas they seemed to spend little time in, though there were no wolves to frighten them off, and so these were ones that quickly developed into woodland. Sallows have light, fluffy seeds that blow on the wind, so they can colonise very quickly. It also seemed that the exact year in which a field came into the scheme affected how it developed – Charlie suspected that it might just depend on whether it happened to be a good mast year for acorns, or ash seeds, or whatever. What was so refreshing about all of this was that it was natural, for nature was taking its course, doing its thing, with no human hand trying to interfere. With no pre-determined outcome, Charlie doesn't mind what happens at Knepp, he is just excited to see what does happen. It took me a while to get my head around this, for all of the conservation efforts I had been involved in before in my life were goal oriented – we wanted to reintroduce an extinct bee, or create 100 hectares of flower-rich habitat, or prevent the spread of an invasive species. Never had I come across the idea that one could just let go, stop trying to be in charge. It was really rather wonderful.

We pulled over by an elaborate, rustic tree platform some six metres in the air, built around a venerable old oak. Charlie shushed me, for I was gabbling noisily to Ted, and we climbed quietly up to the platform. I wasn't quite sure why we were being quiet, but thought better than to ask. From the platform we were looking out over a large shallow expanse of water and marshland. It had once been an arable field, but they had scraped out some of the soil and banked it up to create a shallow lake, doing work that beavers might perform if they were still present. In a similar way, Charlie and his team have also reinstated meanders in the River Adur that runs through the estate. His ancestors had canalised it, creating a deep, straight channel with the intention of draining water effectively from the land during winter floods. There is now good evidence that this just creates worse flooding downstream, but at the time it was well intentioned. It also removes much of the biodiversity. Just as the beavers have done at Yellowstone, restoring the bends creates areas of shallow water, deep water, fast sections and sluggish backwaters, in turn providing a diversity of niches for water plants, insects and fish.

We stood quietly in the tree platform, listening to first a cuckoo and then a turtle dove calling in the distance. Both are species that have undergone huge declines in recent years, but seem to be doing well at Knepp. A red kite cruised silently above us, lazily flicking its long forked tail to steer left and right as it quartered the ground in search of something to eat. Charlie pointed towards the scrub to our right and I noticed three red deer, two hinds and a calf, emerging from the shadows of a sallow copse to wade in the shallows, silently browsing on the tender leaves of semi-submerged willows. Their burnished russet hides glowed in the sunshine, their muscles shivering and their tails flicking to ward off flies. As we watched the deer suddenly froze, alert, their eyes and ears trained on the far bank of the lake where a Tamworth sow emerged from the undergrowth, trailing three piglets behind her. The deer relaxed as the four pigs noisily slurped at the muddy water. I was spellbound – it was almost as if we were on safari in East Africa, but the more remarkable because I was just thirty kilometres from home. On one level I knew these were just pigs and deer – everyday sights – but somehow because they were living wild the whole experience was entrancing.

As we watched, Charlie quietly told me a little more about the pigs. Apparently they sometimes submerge themselves entirely in the lakes, disappearing under water like miniature hippos as they truffle about in the mud of the lake bed in a hunt for swan mussels which they love to crunch up and slurp down whole like huge oysters.[46] In winter they tend to sleep in piles to keep warm. These pigs are new to wild living, but on the whole they seem to have adapted astonishingly well. Watching the sow wallowing in the shallows with her piglets, I couldn't help but be struck by the contrast between these pigs and those reared on factory farms that spend their lives in crates, unable to move more than a foot or two.

---

[46] As a child I once tried keeping one of these magnificent molluscs in a freshwater aquarium in my bedroom, but since they are filter feeders it had little chance of surviving in such a confined space and it inevitably died. Since it lived in the mud at the bottom I didn't immediately realise, until the stench of decay began to seep through my room.

Later in the day we came upon a herd of longhorn cows, half a dozen of them with three calves in tow, wending their way along a network of trails worn between the clumps of bramble and blackthorn, browsing as they went. They have now clearly got over their agoraphobia, and seemed very content with their lot. English longhorns are a charming, scruffily irregular breed, their longish hair blotched blue-grey and white, the colours of an approaching storm, their huge horns curved wonkily. Although at Knepp they live their lives more or less entirely without the interference of man, they remain quite tame, allowing us to approach to within a pace or two. There are public footpaths through the estate, so choosing a placid breed was vital. Apparently breeds with long horns tend to be very gentle since, if they were not, they would be far too dangerous and would long-ago have been culled.

What is most remarkable is that these domestic breeds, maintained in captivity for many hundreds of generations, still retain instincts that can have been no use to them for millennia. Charlie explained how natural behaviours had somehow surfaced in the few years since the cattle were released. They naturally seem to form themselves into herds of a dozen or so, led by a dominant female – if a group becomes too big, it divides into two and they go their separate ways. When she is about to give birth, a cow will leave the herd and go into dense undergrowth in the woods. There she leaves the calf, returning to the herd for much of the day but popping back to suckle the calf at intervals. This might seem like a dubious strategy, for the new-born calf is more or less helpless and she is leaving it unprotected, but this is exactly what many wild animals such as deer naturally do. Perhaps in their evolutionary history, when wolves and other predators roamed the land, tucking the calf out of sight in the woods was a safer bet than having it out in the open, where it would be very obvious to a wolf pack which might then harry the protective herd away from the calf and thus have an easy meal. George Monbiot argues that humans retain ancestral memories or instincts from the time when we were hunter-gatherers, and perhaps these cattle have also inherited behaviours from their extinct aurochs ancestors. I found it reassuring that, even after so long in domestication, caged and with their free will removed, these animals still possess some of the instincts of wild beasts.

Of course the obvious thing that is missing from Knepp is a top predator. In Yellowstone the restoration of wolves has been hugely beneficial, but the sad truth is that Knepp isn't anywhere near big enough to support wolves, even if one could persuade the authorities and the local people that it would be a good idea. Some have mooted the idea of reintroducing wolves in remote parts of Scotland, and this seems plausible (and to my mind is a wonderful prospect), though there is huge opposition and it seems unlikely that it will ever happen. The ecological case is pretty sound – parts of the Highlands suffer from horrendous overgrazing by huge populations of red deer, just as used to be the case in Yellowstone. Wolves actually occur in most European countries, with populations in Spain, Italy, Sweden, Finland and most of Eastern Europe. Small numbers have moved in to France, Germany, even Denmark and the Netherlands in recent years, so is it really so unthinkable to have them in Britain? They pose almost no direct threat to man, despite the lurid headlines,[47] though of course they would take livestock – but if farmers were compensated, is that such a big price to pay? Yellowstone has seen a surge in tourists coming

---

[47] In 2013 the *Daily Mail* reported the sighting of wolves in the Netherlands under the heading, *First killer beast turns up in Holland for 150 years*. I'd take a wild stab in the dark and guess that the *Mail* won't be supporting the reintroduction of wolves to Britain. Shame on them.

to see the wolves, bringing in lots of money to the local area – it seems to me that this might provide far more income to rural communities in remote corners of Britain than they currently obtain from keeping sheep.

Regardless of these arguments, wolves are not practical at Knepp, and nor indeed are lynx, despite being smaller and solitary creatures. Charlie did some back-of-an-envelope calculations and estimated that Knepp could support a population of about half a lynx, such is the large range they need – not likely to form a self-sustaining population. Man has to act as the top predator by culling the large herbivores (unless we are willing to watch some starve in winter), and of course the number they choose to take will influence how Knepp develops, so nature will never be entirely self-willed here. Nonetheless, it feels an awful lot closer to natural than anywhere else I have been in the south of England.

What, you may be wondering, about the humble beaver? Beavers are a native species, and they pose no direct threat to livestock or man.[48] They used to be found all over Europe, but were hunted for their fur and became extinct in Britain in the sixteenth century and in much of Europe by the nineteenth century. Through natural colonisations and no less than twenty-four deliberate reintroductions they are now found once again in much of their former range, though more patchily distributed than before. Evidence that they are overwhelmingly beneficial to ecosystems is clear – by building their lodges, damming streams and excavating water channels they create much new habitat, increasing the diversity of

---

[48] Beavers can carry a nasty tapeworm, *Echinococcus multilocularis*, which can infect people and is fatal, so an obvious precaution we need to take before importing beavers is to ensure that they are tapeworm-free. This is easy enough, as there are plenty of parts of Europe, such as Norway, which have beavers but where the tapeworm is absent.

plants, birds, fish and amphibians. In the USA, the total weight of all the creatures living in beaver ponds is up to five times greater than in the undammed sections. On top of this they are a huge hit with tourists. Why on earth would we not want them back in the UK?

Depressingly, even the reintroduction of this benign, charming rodent was met with much opposition in the UK, and until very recently the UK was more or less the only European country left that did not have beavers. Our National Farmers' Union is strongly opposed to them; in response to a planned release, their spokesman said: 'I haven't seen any evidence that they'll contribute anything to the ecosystem. The history as far as introducing mammals in particular is not a particularly good one. We've seen the grey squirrel, rabbits and even mink so in reality there isn't much evidence to suggest they do any good at all.'[49]

It is hard not to feel embarrassed on behalf of the poor fool who said this. Aside from demonstrating a complete lack of knowledge or simply stubborn denial of the abundant evidence that beavers do greatly benefit ecosystems, he goes on to draw ridiculous parallels with three non-native species. Nobody denies that alien species such as mink can cause devastation, but the beaver is, of course, a native animal. The only reason they aren't here is because we shot them all and turned them into snuggly hats. As well as the NFU, there was also resistance from sporting estates in Scotland who feared that beaver activities might upset the salmon, despite all the studies elsewhere showing that fish overwhelmingly benefit from the activities of beavers.

---

[49] Just as they continue to campaign against any restrictions on their use of neonicotinoid insecticides, a group of highly toxic and environmentally persistent pesticides that appear to be playing a significant role in declines of bees and other farmland wildlife.

Despite the knee-jerk opposition, the argument for a reintroduction eventually won the day, and a small release of sixteen animals took place in 2009 in Knapdale in Argyll, western Scotland. At roughly the same time beavers also appeared as a result of illegal releases in Tayside in the east of Scotland, and in the River Otter in south Devon. The official release was not without its teething problems – some animals were illegally shot by disgruntled locals, and one – a prime candidate for the beaver equivalent of a 'Darwin Award' – felled a tree, which toppled and killed it. Nevertheless, by 2014, fourteen young had been born in the wild, and over 30,000 people had visited Knapdale to see the beavers and their constructions. The release was acclaimed as a great success, and perhaps has helped to persuade authorities in England to allow the Devon beaver population to remain. This has now become an official reintroduction programme, with formal monitoring by the Devon Wildlife Trust. At the last count there were at least twelve animals, including one young wild-bred kit born in the summer of 2015. Knepp almost certainly has enough freshwater habitat along with plenty of young trees for them to gnaw and fell, and it would be fantastic to see how their activities help to shape the landscape. Charlie has hopes that this might one day become a reality.

After a day at Knepp, appetite whetted, I was keen to find an excuse to return. Luckily, just such an excuse soon presented itself. Charlie and the team at Knepp are very keen to find out more about what wildlife they have, and how it is changing over time. They've had quite a number of surveys, for example of the plants, birds and butterflies, but no one had surveyed the bees, and Charlie asked me if I'd be willing to do it. It was too good an opportunity to miss – the chance to wander around Knepp once a month through the spring and summer was irresistible. I'm pretty good at identifying bumblebees – it would be embarrassing if I were not after all these years of studying them – but I'm far from fully competent in the identification of some of our smaller, solitary bee species and so I asked my PhD student Tom Wood if he'd like to come along. Tom is a classic obsessive entomologist – although I guess you might be thinking about pots and kettles at this point. His PhD is on looking at how effective wildflower strips on farms are at increasing bumblebee populations, so he spends all his working days studying bees. At weekends, he goes looking for bees – in particular the rarer solitary species. His holidays are spent in southern Europe looking for more bees. Come his birthday, he requests that his presents be obscure books on bees, rare and expensive tomes with gripping titles such as *A Key to the Halictidae of South-western Moldavia.*[50] You might, quite reasonably, have gained the impression that I am a bee fanatic – but next to Tom my interest seems almost casual, a take-it-or-leave-it, amateurish kind of hobby. Tom is the real deal, perhaps one of the last of a dying breed; just the man I needed if I was to do a thorough job of discovering all the bees at Knepp.

So it was that in the middle of April 2015 we began our bee survey. Penny Green, the Knepp ecologist, joined us, keen to learn more about bees. It was a slightly overcast, chilly day, and we did not get off to a great start. In early spring there are not too many flowers at Knepp – much of the turf is short at this time of year, grazed down over the winter by the animals, and there were few flowers save the blackthorn hedges and scrub which were in full,

---

[50] This isn't actually a real title, should you be tempted to rush out and buy a copy, but some genuine ones are scarcely less obscure. If you're in the market for an identification to British bees, I would strongly recommend the excellent *Field Guide to the Bees of Great Britain and Ireland* by Steven Falk, with lovely illustrations by the renowned wildlife artist Richard Lewington.

dazzling bloom. Unfortunately, blackthorn isn't very popular with bees, so we struggled to find many of them, just a handful of queens of common species such as buff-tails and red-tails.

Perhaps more exciting were the other creatures we saw. Penny had scattered pieces of old tin around the estate, which did look untidy and are of course rather unnatural, but they provide a great way to monitor reptile populations. Snakes and lizards tend to gather beneath them, particularly on cool days in spring, for the tin absorbs heat and allows them to warm up. Under the very first tin I lifted was a knot of slow-worms, ranging in colour from dark brown to silver to fawn, a knot that untangled itself as the animals glided slowly away in silent panic at my intrusion. The next tin had more slow-worms, and a goodly sized grass snake, perhaps three feet long. The next, close to the scrape where Charlie and I had watched the pigs wallow, had a couple of common frogs and both a smooth and a great crested newt. I could have spent the rest of the day on a herpetological adventure turning over bits of tin, but this not being a prime way to find bees I thought I'd better focus on the job in hand.

Near the scrape was a stand of young sallows in full bloom, and it was here that we found most of the bees. Sallow blossom is an absolute favourite with bees, the yellow powder-puff flowers of the male trees producing both pollen and nectar, the more drab green flowers of the females producing just abundant nectar. Given that there are rather few other common flowers in bloom in April, these trees are almost certainly a mainstay of the diets of many queen bumblebees, as well as some of our solitary bee species. It is just as well that sallows are doing so well at Knepp. Queens of several of the common bumblebees were there, mainly early and buff-tailed bumblebees, plus a few early workers.

There are some small areas of ancient woodland at Knepp, and it occurred to me that this could provide a good place to find bees.

It was just about time for the bluebells to be in flower, and in a good patch of bluebells one can usually find a few bumblebee queens foraging. Bluebell woods are an iconic British habitat – although bluebells have a wide distribution, nowhere else do they occur at such high density as in Britain, where they form swathes of blue in mid to late April, before the leaves of the trees above burst out and shade them for the rest of the year. I headed over to a stand of large oaks to see what was beneath them, but instead of bees or bluebells I found pigs. Three sows and innumerable piglets were lying in a dusty heap amidst a scene of devastation. The woodland floor had been thoroughly rootled. Charlie was later to tell me that pigs love to eat bluebell bulbs, and so for them a bluebell wood means one thing – meal time. On the Continent, woodland flora is more mixed and diverse, but perhaps not as spectacularly beautiful. A few bluebells remained, blooming amongst the mounds of earth, but most had gone. Wood anemone, wild daffodils and primroses were creeping in instead, and I guess they will spread over time.

This raises an interesting question. Are our beautiful bluebell woods an artificial habitat, only possible in Britain because our overzealous hunting of wild boar accidentally exterminated the bluebell's biggest enemy? Are those of us that would like to see wild boar living across the UK in the wild,[51] as they still do in

---

[51] There are actually quite a few feral 'wild' boar living in the wild in the UK, escapees from farms – they are masters of escapology. There are small populations in the West Country, East Sussex and perhaps as many as 500 in the Forest of Dean. There has been great controversy over their repeated culling, justified on the grounds that they damage trees (last time I looked, trees seemed to cope elsewhere in Europe) and that they dig up amenity grasslands (in my view perhaps no bad thing). Don't wild boar have as much right to live here as we do?

almost all of Europe, willing to pay the price that we may lose many of our bluebell woods? I would say yes, but that of course is just my value judgement, and others will no doubt disagree. There are often no right answers in conservation.

I snuck up close to the sleeping pigs, camera in hand and snapping away, until suddenly they detected my presence and exploded in a cloud of dust, grunts and squeals. It was hard to know how to react to such huge animals at such close quarters, and I was momentarily aware of just how much bigger the sows were than me. Were these wild animals or domestic ones? I wasn't quite sure, and a pulse of adrenalin surged through me. Of course, once the pigs got over their surprise they were fine, and the adults settled back to their slumber, while the inquisitive piglets crept up to me to have their portraits taken.

I realised once again that I had digressed from the mission to find bees, so I left the pigs to it, and went in search of Tom and Penny. In the meantime, Tom had been busy searching for solitary bees, most of which are rather small and difficult to identify without a lot of practice. They don't even have a satisfactory name as a group, for although they are usually referred to as solitary bees some are not in fact solitary. Tom prefers the technically accurate 'non-corbiculate' bees, the corbicula being the proper name for the pollen basket possessed by bumblebees and honeybees but not by these other species, yet this is an abstruse term that would mean nothing to most people. Perhaps the term 'forgotten bees' is best, for these obscure creatures are little studied by scientists and almost entirely unknown to most of us. Yet what we do know about them suggests that they are important in pollinating crops and wildflowers, and that they have fascinating, complex and highly variable life histories.

Many of these forgotten bees tend to nest in aggregations, clusters of sometimes hundreds or even thousands of burrows in the ground, each marked by a small conical volcano of excavated earth. In most species, each female has her own burrow (though how she remembers which is hers is a bit of a mystery), single-handedly stocking it with pollen and sealing her eggs inside, where they develop without further attention. A few of these 'solitary' bees are truly social, with the queen founding a nest and then rearing a batch of workers who then help her to rear a batch of males and new queens – in other words their life cycle is very similar to that of bumblebees. When I finally caught up with Tom he had found a nesting aggregation of one such species, the less than snappily named sharp-shouldered furrow bee,[52] or *Lasioglossum malachurum*, a greyish bee just six millimetres long. The nests were scattered along the bare, packed clay of a vehicle track. Males were patrolling busily backwards and forwards above the nest entrances, hoping to mate, while the females came and went, bringing back pollen loads and doing their best to avoid the attentions of the amorous males.

Instead of looking for bumblebees on the blackthorn, which had proved to be pretty unproductive, I started scanning bare patches of earth for bee nests. In no time at all we'd caught specimens of no less than ten different species of bee, most of them members of a genus called *Andrena*, mostly smallish bees, often with shiny dark abdomens and commonly known as mining bees. Globally, there are no less than 1,300 known species of mining bee alone, most of them very easily overlooked. Some were rather handsome – such as the orange-tailed mining bee, *Andrena*

---

[52] Until recently, most of our 'solitary' bees had no common name, but in an attempt to encourage people to show more interest in them, and to make them seem more accessible to the amateur, common names have been invented for them. This isn't everyone's cup of tea – Tom is a purist, scornful of the need for such dumbing down.

haemorrhoa, which has a fox-red thorax and bottom, and *Andrena fulva*, one of the few solitary bees that is eye-catching enough to have long been accorded a common name: the tawny mining bee. When I was a child we had an aggregation of these large, rust-coloured bees nesting in our lawn, and I loved to peer down into their holes and see the females staring back up at me. We also found a little nomad bee, *Nomada flavoguttata*, a minute, almost hairless, reddish bee that is a 'cleptoparasite' – it specialises in nipping in to the *Andrena* nests when the mother is away and laying its own eggs on the pollen stores therein. Nomad bees are very like cuckoos – their eggs hatch fast, and the very first thing the newly hatched nomad bee grub does is to kill the egg or grub of the host. The young nomad larvae are equipped with large, sickle-shaped mandibles expressly for this purpose. Once the host's offspring has been dispatched, the nomad grub can consume the pollen store at its leisure. As it grows and moults, it loses the murderous mandibles, which are of no help in consuming pollen. This might seem a pretty underhand strategy, but clearly crime pays because there are about 850 species of nomad bee known in the world, each adapted to a different host.

Over the next few months, on our successive visits, it was fascinating to see how the vegetation changed through the season. In early spring, the turf had been close-cropped, leaving few flowers. Over the winter the plants would have been able to grow little, and so were grazed right down by the cows, deer and horses. If that had remained the case through the spring and summer, as I had suspected it might, then we would have found few bees. What I hadn't reckoned on was the fairly obvious point that plant growth accelerates through the season, while the number of animals remains more or less the same. By late May the meadows were a sea of creeping buttercup, sprinkled with patches of blue speedwell with purple ground ivy spreading in the shadier spots and the first bramble blossoms appearing in the thickets. By July, the open areas were knee-deep in white clover, the tiny mauve blossom of smooth tare, with clumps of red clover and bird's-foot trefoil. Grazers love to eat clover, but there is clearly far more of it than they can consume, and so it gets to flower. In August, the common fleabane takes over, producing yellow daisy-like flowers on grey-green, woolly stems. This plant is unpalatable to grazers, and has become very common in much of the pasture at Knepp. Charlie might have been tempted to control it, because it reduces the fodder available for the livestock, but of course that would be contrary to the philosophy of what he is doing. If fleabane runs amok, then so be it. My guess is that some balance will soon be restored, as types of insects that are able to eat fleabane arrive and take advantage of the abundant food source. Time will tell.

Through the year we clocked up a list of ten bumblebee species, none of them particularly rare. Much more exciting in many ways were the other bees, of which we found forty-two different types by the end of the summer, including many more mining bees and a few more types of nomad bee, plus yellow-faced bees, sweat bees, mason bees and more. Some are very scarce bees on a national scale. Perhaps most thrilling of all was finding another cleptoparasite, the rough-backed blood bee (*Sphecodes scabricollis*), which is exceedingly rare across Europe and in the UK is listed as a 'Red Data Book' species, making it a high priority for conservation. Blood bees differ from nomad bees in that the adult female enters the host nest and kills the host's offspring herself before laying her own egg, rather than leaving her newborn offspring to carry out the dirty deed. This particular species is rather tiny, at just six millimetres long, black with a bright red band around its abdomen, and it specialises in attacking the nests of the flamboyantly named bull-headed furrow bee (*Lasioglossum zonulum*).

Of course I am sure that we missed many more. Searching for creatures which may be just five millimetres long in a 3,500-acre estate, it is reasonable to assume that one won't find them all. We also don't know which ones were here before the rewilding began. In an ideal world, Charlie would have had the entire estate subject to a thorough inventory of its wildlife while it was still a regular farm, and then we could see how it was changing over time, but he had neither the time nor the resources. At least we will be able to see how it changes into the future – which new species arrive, and which ones disappear. It will be fascinating to watch. Perhaps some of the rarer bumblebees will eventually arrive. I suspect that the flora will become steadily more diverse as time goes on – it may take many decades for some species to get here. What is certain is that the flora and fauna of Knepp will change over time, following natural processes.

It is interesting that humans are so averse to change, so keen to maintain the status quo. Charlie's project had met with a lot of resistance at the outset – some locals felt that it was immoral to abandon farmland, that it was a farmer's duty to keep the land neat, tidy and productive. Of course this is a dubious contention; we may have got used to modern, large fields containing neat monocultures but they are a relatively new phenomenon. One hundred years ago, before mechanisation and the advent of chemical herbicides, farms would have been much messier, a patchwork of small fields separated by dense hedges, some fields with crops and colourful arable weeds, some left fallow with yet more weeds, some kept for hay, others grazed by livestock. Four thousand years ago it was probably mostly forest. Four million years ago it was forest or perhaps something savannah-like, roamed by elephants. Who is to say what the land should look like? But humans instinctively resist change, and want to hang on to whatever they are used to, however humdrum.

Early on in the project, one particular family, who live in a cottage surrounded on three sides by the new wilderness, started complaining that they were being kept awake at night by birds singing. A little investigation revealed that the racket was due to a medium-sized, rather shy brown bird that had taken up residence in a nearby thicket – a nightingale. Knepp had previously had no nightingales, but it now supports up to 32 singing males (and an unknown number of breeding females) each year. Nightingales like to nest a foot or two above the ground in dense, spreading thickets, and as the abandoned hedges at Knepp spread sideways they have come to provide perfect habitat. The males arrive back from their overwintering grounds in Africa in late April, and they find a good nesting site and then sing to call down the females as they fly north. Since nightingales migrate mostly at night, that is when the males sing, serenading the females as they pass over in the dark, hoping to lure one down. Their song is of course famously beautiful, a complex mix of liquid trills and warbles[53] – presumably because female nightingales are pretty hard to impress, and hence the males must sing their hearts out to stand any chance of wooing one. So, imagine the hardship suffered by Charlie's neighbours, having to listen to the awful din of nightingales every night – one's heart bleeds. Encouragingly, the story ended well. These folk now know what the noise is, and eventually came to appreciate it, even to love it, and have realised how privileged they are. As I said, we humans can be oddly resistant to change, no matter how benign.

The Knepp project is only really possible because of taxpayers' money, in the form of an annual subsidy in the region of a quarter

---

[53] Though the song of the nightingale is admittedly impressive, I personally think that you cannot better the song of a blackbird on an early spring morning.

of a million pounds. Other income comes from the sale of high quality meat, 'safaris' that are run to see the wildlife, a 'glamping' site,[54] hosting weddings in fancy tepees, renting out former farm buildings as offices and workshops, and anything else Charlie and his team can dream up. Overall, they are running a small profit, enough to keep them afloat. You might question why your taxes should pay for this? Personally I think it is a bargain, particularly when compared to many of the other things on which our money is spent,[55] though by now you'll have guessed that I have bought the concept hook, line and sinker.

What is perhaps most fascinating for me about Knepp is that its most striking successes were never planned or anticipated. There was no deliberate intention of encouraging nightingales, yet they arrived. Nightingales have undergone a drastic population collapse in the UK – between 1967 and 2007 the UK nightingale population

---

[54] 'Glamorous camping' in an eclectic collection of rather comfortable tents and shepherd's huts.

[55] There is a fascinating website, farmsubsidy.org, which tells you exactly how much subsidy every 'farmer' in the EU gets each year. You'll be perplexed to hear that the sugar producers Tate & Lyle are the biggest UK recipient, having received no less than €594,270,084 of 'farming subsidy' over the last fifteen years (yes, that is the correct number of digits). They don't actually do any farming whatsoever – they simply buy sugar cane from tropical countries and process it into refined sugar. We also give similarly astronomic sums to companies who grow and process sugar beet, a crop that is grown in intensive monocultures with the application of many pesticides, all to produce a substance that is fundamentally bad for us, and which is one of the main contributors to the epidemic of diabetes that is helping to cripple our National Health Service. Perhaps this is not the wisest expenditure of taxpayers' money.

fell by 91 per cent, the biggest fall in numbers of any UK breeding bird since records began. If one had set out to create habitat for it, what Charlie has done at Knepp would almost certainly not have been seen as the way forward – the decline of nightingales has been blamed largely on overgrazing by the UK's growing deer population, which is thought to have damaged the thickets in which it nests – so creating a large enclosure full of deer and other grazers would not have seemed like a great idea.

Another great success story at Knepp has been the rise of the purple emperor. Glance through a guide to British butterflies, and two species stand out as being especially spectacular: the swallowtail, a majestic yellow and black kite-shaped butterfly with elegant tail streamers, sadly found only on the Norfolk Broads; and the purple emperor, a big, powerful insect, the males with iridescent purple wings. Both were species that I could only have dreamed of seeing as a child in rural Shropshire. The purple emperor is an elusive beast that is generally found in large tracts of mature, deciduous woodland in the south of England. Even in such places it is rarely seen, for the butterflies spend most of their day in the treetops; they feed on honeydew, the sugary droppings of greenfly, so have no need to come down to flowers. The males set up territories in the canopy of an especially tall tree, known amongst emperor aficionados as the 'master tree', and there they engage in aerial skirmishes not unlike those of *Bombus hortulanus*, their wings glinting in the July sunshine. The females lay their eggs on sallow trees growing along the edge of woodland rides or clearings. If one wished to conserve purple emperors, one would probably conclude that the best strategy would be to protect ancient woodlands at all cost. If anyone had suggested that arable farmland could become prime emperor habitat within just ten years, they would have been regarded as nuts. Yet this is exactly what has happened at Knepp. The sallows have proliferated in some of the meadows, and the standard oaks along the overgrowing hedge-lines provide master trees. Purple emperors were simply not found at Knepp before the rewilding began; by 2013, no less than eighty-four males were seen in a single day (surveys tend to focus on the territorial males which can be counted through binoculars, a tricky business but much easier than trying to spot and count the less colourful and less active females).

Sadly, a recurring feature of man's activities is that we endlessly implement change – sometimes deliberately, often accidentally. For millions of years, change tended to be very gradual on this planet. Aside from the occasional asteroid strike, tens of millions of years would go by without anything particularly much happening. Ice ages came and went over periods of thousands or tens of thousands of years. Each year a handful of species naturally went extinct, but new species gradually evolved so that there has been a net increase in global biodiversity over the millennia – until very recently. Nowadays, large-scale, man-induced changes can occur in years, sometimes in just hours. We clear forests for farming, introduce invasive species, introduce set-aside schemes in farming, decimate fish stocks, plant dense woodlands of non-native conifers and then cut them down, drain marshes, create dams and reservoirs, abandon marginal lands, scrap the set-aside schemes we'd introduced just a few years earlier, cause acid rain then partly fix the problem, make holes in the ozone layer then partly remedy that too, alter the climate, exterminate large predators, introduce endless new pesticides and other pollutants and then ban some of them only after we have seen the predictable damage they do – a ceaseless barrage of change in the face of which wildlife has to either adapt or die. Most creatures can adapt to gradual change, particularly if they start with a large, genetically diverse population, but few can cope with the continual, rapid changes that we throw at them. I sometimes wish we could just learn to stand still for a while, to let nature catch up, but that seems to be the one thing that mankind is most unlikely to do any time soon.

Knepp is a wonderful example of what can happen if we just stop. Stop doing anything at all – stop trying to conserve, stop trying to interfere, stop trying to manage – just let nature catch up, absorb the changes that went before, and then do its own thing. Who knows what will happen next at Knepp, what new species will arrive? What will it look like in a year, a decade, a hundred years? The truth is that we really don't know, we cannot anticipate – and although I am a scientist, and it is my job to seek to understand and predict these changes, I rather like that.

Is it too much to think that we might have a rewilding project in every county in Britain? Places within reach of all of us, where we could experience a sense of adventure, of nature unchained. Where we could hope to glimpse the purple flash of an emperor, hear a nightingale sing, perhaps hear the slap of a beaver's tail as it warned its family of our presence.

I'm not suggesting that we abandon traditional nature reserves – they have an important role to play – but it is clear that the traditional model of conservation, setting up reserves to protect a particular rare habitat or species, has not managed to stem the overwhelming tide of wildlife decline. Perhaps there is another way, one that might help us to reconnect with the natural world.

Conservation can be a depressing subject. It often feels as if we are fighting a rear-guard action, a losing battle against the relentless tide of human population growth and the futile, senseless drive for economic growth at all costs. If you ever feel that all hope is lost, then go to Knepp, or Canvey Wick, and take heart. Nature is fantastically resilient, and it will recover, though of course the more damage we do the longer that will take. Is there anything closer to true magic in this world than the transformation of a lagoon of grey, industrial ash into a flower- and insect-filled meadow?

There will one day come a time when we stop messing up the Earth – either because we have wiped ourselves out or because we have learned to live amongst nature, rather than trying to exert dominion over it. When that happens, wildlife will come back, creeping from the cracks in the concrete, sprouting from the seeds that remain in the soil, adapting, thriving, evolving into new and wonderful forms. It would just be nice if we or our children were here to see it.

Epilogue: Back-garden Bees

The world's biodiversity is under threat, particularly in such exotic places as Ecuador, a country blessed with extraordinary natural riches but also with an impoverished and rapidly growing human population. It is easy for us to criticise from afar, bemoaning the deforestation, the pollution and the foolish deliberate introductions of invasive species, while living in our comfortable, warm houses, watching our flat-screen TV from Korea while snacking on almonds from California, sipping an Australian Shiraz, and pondering whether to splash out on a holiday to the Maldives or stick with Tenerife again. In truth, we in the developed world are in no position to preach, for we have already devastated our own countries, stripping them of forests long ago, scrubbing clean the land of most of its wildlife to create cities, motorways, shopping centres, golf courses and of course vast monocultures of crops. Even when we discover hotspots for wildlife that have somehow sprung up right under our noses, in our biggest cities, we do a poor job of protecting them. We create the demand for much of what happens in the developing world, with our endlessly increasing consumption of fossil fuel, food, minerals and other resources. It is often huge companies based in Europe and the USA that buy up cheap land in developing countries and impose industrial farming regimes, or conduct devastating mining activities that rip great holes in the land and pollute rivers and soils. In any case, we can hardly blame people in developing countries for trying to attain our luxurious lifestyle.

We could afford to save the world, if we so chose, and surely we must. Skipping one can of Coke in five, or an equivalent, would hardly place us in great hardship. We could rein in the blatant profiteering of multinational companies who take advantage of the weaker environmental legislation in developing countries. We could pay poorer countries to protect their wildlife, and we would barely notice the cost. But we also need to put our own house in order, for we too continue to prioritise new rail links over preserving ancient woodland, we pollute our farmland with a blizzard of chemicals, and we spend millions prospecting for gas in shale rocks while pretending to care about tackling climate change.

All of these issues are hard to fix. We often feel helpless. The main political parties rarely mention the environment except as a token gesture in the run-up to an election. Remember David Cameron's quick trip to the Arctic to hug a husky during the 2010 election campaign, in which he promised the 'greenest government ever', before appointing the climate-change sceptic Owen Paterson as Environment Secretary? But of course we are not entirely helpless. Aside from voting every few years, we make important decisions every day. As Jane Goodall said, *We have to realise that each day we make some kind of impact. And we have a choice as to what type of impact we will make.*

Conservation begins at home. We should all recycle everything we can – it is possible to almost entirely avoid producing waste that has to go to landfill, especially if one avoids buying food with unnecessary or unrecyclable packaging. I recently spotted on social media a caption someone had written beneath a picture of rows of bananas on sale in Morrisons, where each individual banana had been packed on a polystyrene tray and covered in cling film.

The comment read, *If only bananas had evolved with their own hygienic removable wrapper.* Why do we put up with such nonsense? We should all try to buy locally produced food, ideally from organic farms. There are plenty around, and they need our support. We could all do without buying strawberries from Chile in January, and perhaps if none of us had bought them the Chilean government might not have felt the need to import our bumblebees. If customers demand it, supermarkets will soon respond, and our collective buying power could have a huge influence on the way we produce food at a global scale. We should all have a compost heap or a wormery, or both, though this might be tricky if you live in a flat. If you have space, have a go at growing your own, healthy, nutritious food, and plant flowers to encourage bees, butterflies and birds. It might not seem like it, but every small decision makes a difference – after all, there are now more than seven billion of us, and we each hold the future of our planet in our hands.

Imagine if every garden in Britain was wildlife friendly, with cottage-garden herbs and wildflowers, healthy home-grown veg and perhaps a home-made bee hotel for solitary bees to nest in in the corner. Why don't we ban pesticides in our gardens and urban areas? Quite a few cities around the world have done this, and they are not overrun with pests. Imagine also, council-owned land managed to encourage wildlife: road verges and roundabouts not mown every five minutes, but instead sown with wildflowers; grassy areas in parks allowed to grow long in places. Let's persuade our local authorities to stop putting out annual bedding plants every spring, which are no use to wildlife, but instead to plant the borders in our parks with bee- and butterfly-friendly perennials. What about having patches of flower-filled hay-meadows on university campuses and in school grounds. Perhaps our industrial estates and science parks could be planted with native flowering shrubs that provide food for bees and berries for birds, rather than planting evergreen exotics. Why don't we plant apple, pear and plum trees along our suburban avenues so that the residents could pick fruit along the street, and children could pluck an apple on the way to school? We might sprinkle in some green roofs and green walls on new buildings. Perhaps we could protect wildlife-rich brownfield sites and open them up to the public, rather than allowing them to be tarmacked over. We could green our cities, encourage wildlife in to live amongst us, and create the largest nature reserves in Britain, all for no net cost whatsoever. Our children could grow up connected to and respecting nature, able to catch grasshoppers in their hands amongst the long grass, watch the bees bustling amongst the runner bean flowers, or hunt for newts and great diving beetles in the local canal. If this is what we want for them, then now is the time to act. My fervent hope is that future generations will have the chance to experience the natural world first hand, so that they too can fall in love with it. One of my greatest fears is that my grandchildren, should I ever have them, will grow up in a grey, impoverished world of concrete and steel, unable to experience nature for themselves because it has all but gone, and not knowing or caring because they have no idea what they have missed. It doesn't have to be this way. It is within our grasp to green our cities. Increasing urbanisation is inevitable, so let's use our imagination to make our urban areas into sprawling nature reserves, where people and wildlife live alongside one another in harmony. Perhaps it is too fanciful to think that our cities might become 'Britain's rainforests', but our children might thank us if we try.

Index

penguin.co.uk/vintage